Computers in Geography

PAUL M. MATHER

Computers in Geography

A PRACTICAL APPROACH

BASIL BLACKWELL · OXFORD

ISBN 0 631 16870 2

Filmset and printed by
BAS Printers Limited, Wallop, Hampshire

Contents

Preface

By the end of this decade, the geographical profession will recognize that the ability to understand and use a computer is as important as the ability to understand and use quantitative methods. Over the last ten or fifteen years the use of computers in geography has increased as the complexity and sophistication of geographical methods has grown. Along with the now almost traditional statistical and data-processing techniques, such methods as automated cartography, information storage and retrieval, and simulation have also been introduced into geography. These techniques share one requirement—the need for a computer to carry out the enormous number of calculations.

In view of the foregoing, it seems paradoxical that, at the time of writing (1975), no textbook had yet been written to introduce the geographer to modern computing methods. Some geography departments do offer their students a course in 'computer appreciation', but such courses are often given by members of staff from the local computing centre who are unable to provide the student with examples of the geographical relevance of the material they present. In this respect, this situation closely resembles that which followed the so-called quantitative revolution in geography, when statistical methods were largely taught by non-geographers until the geographical profession developed the skill and expertise necessary to provide practical courses in the geographical appreciation of statistical methods. Textbooks such as Gregory (1963) helped to bring about this change. The fact that students could be shown the relevance of statistical techniques to geographical problems provided a greater degree of motivation than could be aroused by the non-geographer who had, of necessity, to rely either on trivial geographical examples or else on more realistic examples drawn from his own subject. The appreciation of these non-geographical examples often presented a greater problem than the understanding of the technique that was being demonstrated.

The same arguments and observations could be applied to the present-day teaching of computing in geography departments. The only textbooks available are written for electrical engineers, chemists, economists or mathematicians. The geography undergraduate may spend more time in attempting to unravel the mysteries of the laws of mechanics or of differential equations than in learning about computers. The present book is an attempt to fill an obvious gap by providing a course of instruction in both the use and the application of computers in geography. The first two chapters consist of background material. The third chapter is presented in the form of sections; each section introduces a new idea and provides several examples and exercises which are of a geographical nature whenever possible. This chapter could provide the basis for a series of practical classes in computer programming, or of a course of self-instruction. The fourth chapter consists of a series of larger scale examples which show the usefulness of the computer in various geographical situations, involving statistical analysis, simulation, and description of point patterns. This material could be used in tutorial projects, or as follow-up work for the student.

Acknowledgements are due to the following companies, which provided illustrative material: Calcomp Ltd., International Business Machines (United Kingdom) Ltd., and International Computers Limited.

The author wishes to thank Dr. J. P. Cole, of the Department of Geography, University of Nottingham, who has encouraged the writing of this book and has provided advice, examples and much constructive criticism. Any lapses are entirely the author's responsibility. The manuscript was ably and efficiently typed by Miss Anne Limbrey.

Paul M. Mather,
Nottingham, January, 1975.

Notes
(1) Since some of the computer-printed lines are wider than the pages of this book, the symbol '↔' has been used at the end and beginning of successive lines to indicate that they should form a single line.
(2) The symbol 'Δ' has been used to indicate a space.

List of Figures

List of Tables

List of Tables

CHAPTER 1

Introduction

Geography has been defined as the study of the earth as the home of man. Most other academic disciplines have also been succinctly defined. Mathematics, for example, has been called the study of numerical logic, and physics the study of the properties of matter. Such definitions are necessary in order to delimit a particular field of enquiry, but their disadvantage is that they stress the uniqueness and independence of each discipline and fail to point to the linkages between disciplines that are an important part of science as a whole. One such link is that connecting geography and mathematics (Cole, 1969). Parts of geography and certain branches of mathematics have had close connections for several centuries, as for example, the study of map projections and geometry. Geography and mathematics have however become much more closely linked since the 1950s. For example, the use of statistical methods is now widespread, especially in urban social geography, geomorphology and hydrology. The study of networks has become an important aspect of human geography (Haggett, 1965; Haggett and Chorley, 1969), while advances in automated cartography have resulted from an increasing interest among geographers and geologists in the problem of spatial interpolation. One of the most radical changes in geographical research and teaching in recent years has been the introduction and use of simulation techniques, which allow the geographer to experiment with 'alternative realities'. (Cole, 1975.)

All these developments in geographical methodology have one requirement in common—the need for a digital computer to carry out complex and lengthy calculations. Early users of numerical methods in geography could handle the relatively simple calculations facing them with the aid of a hand-calculating machine, and teaching methods were based on formulae that were suitable for such machines (Gregory, 1963). However, the fact that most colleges and universities now have access to a digital computer means that the time consuming job of computing can be given to a machine, thus leaving the student or the research worker free to concentrate on the meaning of his results, rather than on the arithmetical operations required to produce them.

Although geographers had few inhibitions about the use of hand-calculating machines, there has been passive, though not necessarily active, resistance to the use of the computer. This is due, in the main, to a belief that computers, like animals in a zoo, need experts to look after them. Geographers have, therefore, tended to rely on others to write computer programs or to suggest how computers could be used in geographical work. This approach has one major drawback. Computer experts are rarely well-informed about geographical problems, while geographers are frequently unaware of the capabilities and limitations of computers. It is difficult to see how much meaningful dialogue could develop in such a situation. The first aim of this book is to provide the geographical reader with a background knowledge of computers and the way they operate. Such knowledge may be useful in a practical situation. In addition it has been argued that all students in higher education should have a basic knowledge of computers, since both in everyday life and in many areas of academic study computers are assuming an increasingly important role. The second aim of this book is to present a course of instruction in computer programming, using the Fortran language. The obvious reason for learning to program is to gain sufficient skill and experience to be able to make use of a computer. However, there is another reason. Programming a computer requires a combination of patience, intelligence and logical thought. It would not be difficult to support on academic grounds alone the argument that a course of instruction in computer programming should be an essential part of a university or college course. In time, perhaps, a computer language such as Fortran might take the place once occupied by Latin in the school syllabus.

The present book was thus begun with two principal aims: (i) to provide some basic facts about computers, the way they operate, their abilities and limitations, and (ii) to present a short course in Fortran programming for the first or second year geography student. Chapter 2 contains an introduction to computers, with special emphasis on their structure and mode of operation. Chapter 3 consists of a course in Fortran programming. The basic operations of input, output and computer arithmetic are discussed in this chapter. Teaching is mainly by example, though summaries are provided for reference

Figure 1.1

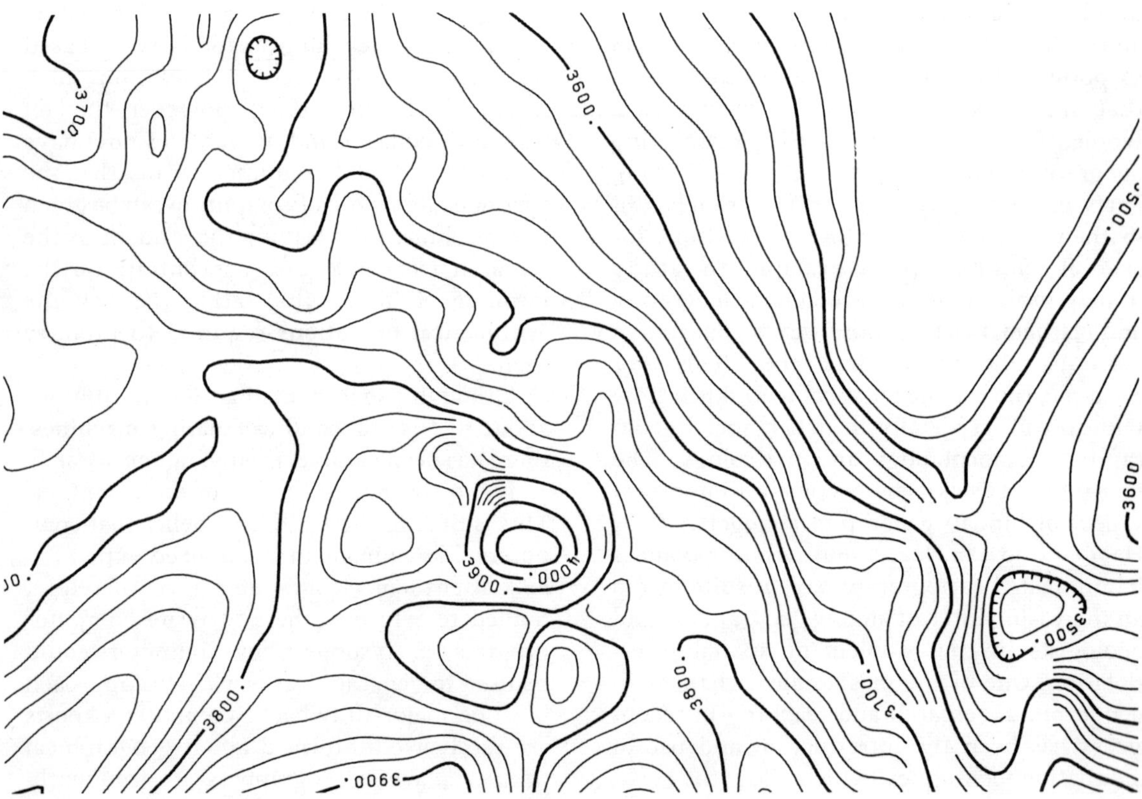

purposes. The exercises which are included in each section are an essential part of the text, and answers to selected exercises are given in Appendix D.

The fourth chapter contains five case-studies illustrating the use of the computer in geographical situations. These case studies are intended to demonstrate the way in which a computational problem is approached. It is often more difficult to work out the strategy to be adopted in a particular problem than it is to write the computer program itself. Chapter 4 also includes further details of the Fortran language of a more advanced nature than that presented in Chapter 3.

No mathematical skills are assumed, nor is any knowledge of advanced mathematics necessary. In writing this book, however, it has been taken for granted that the reader has some knowledge of elementary statistics to the level of Cole and King (1968), Ebdon (1976) or Gregory (1963).

To end this chapter, some examples of the use of computers in geography are illustrated. Figure 1.1 shows a contour map produced by a computer-driven graph-plotter. Such plotters are highly accurate and can reproduce exactly a given base map as many times as is necessary. The graph-plotter can also be used to draw perspective views of an area from a specified point. An example is given in Figure 1.2. Statistical calculations are perhaps those most frequently required by the geographer; the example given in Figure 1.3 (see p. 4) took less than 1 second of computing time on the Nottingham University ICL 1906A computer.

Figure 1.2

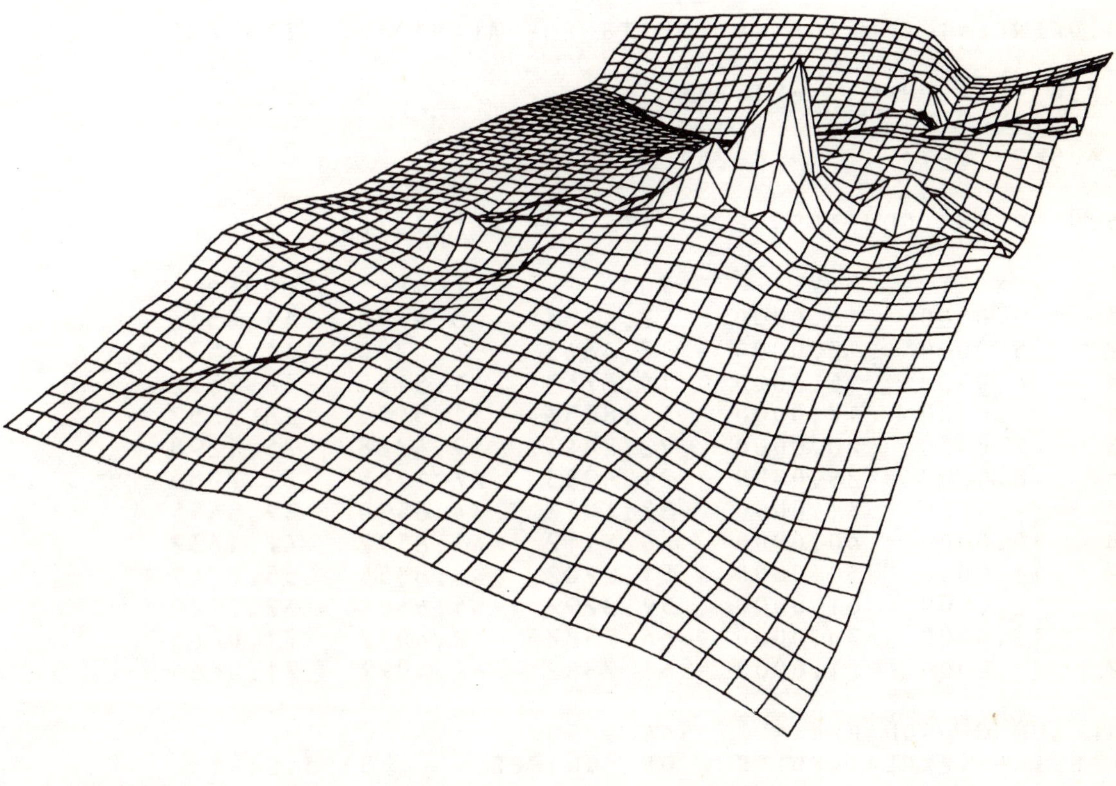

Figure 1.3

```
REGRESSION  EQUATION  FOR  THIS  DATA IS Y =  -29.579 + 7.409X

     X          Y      PREDICTED Y    RESIDUAL

   5.000     10.000       7.466         2.534
   5.000     12.000       7.466         4.534
   6.000     15.000      14.875         0.125
   6.000     17.000      14.875         2.125
   7.000     20.000      22.284        -2.284
   8.000     26.000      29.693        -3.693
   9.000     31.000      37.102        -6.102
  10.000     40.000      44.511        -4.511
  11.000     50.000      51.920        -1.920
  12.000     61.000      59.329         1.671
  13.000     70.000      66.738         3.262
  13.000     71.000      66.738         4.262

EST. STANDARD ERROR OF ALPHA =     3.483
EST. STANDARD ERROR OF BETA  =     0.378
VARIANCE OF RESIDUALS =   14.323

  95.0PERCENT CONFIDENCE LIMITS FOR ALPHA AND BETA ARE

  -37.339   ALPHA     -21.818

    6.567   BETA        8.251

  95.0PERCENT CONFIDENCE LIMITS FOR PREDICTED Y ARE

          X           Y          YBAR        LOWER      UPPER
   1    5.0000     10.0000      7.4663      3.4790    11.4536
   2    5.0000     12.0000      7.4663      3.4790    11.4536
   3    6.0000     15.0000     14.8753     11.5155    18.2352
   4    6.0000     17.0000     14.8753     11.5155    18.2352
   5    7.0000     20.0000     22.2843     19.4388    25.1298
   6    8.0000     26.0000     29.6933     27.1785    32.2080
   7    9.0000     31.0000     37.1022     34.6590    39.5455
   8   10.0000     40.0000     44.5112     41.8592    47.1632
   9   11.0000     50.0000     51.9202     48.8355    55.0049
  10   12.0000     61.0000     59.3292     55.6664    62.9920
  11   13.0000     70.0000     66.7382     62.4097    71.0666
  12   13.0000     71.0000     66.7382     62.4097    71.0666

TOTAL SUM OF SQUARES =     5646.2500
REGRESSION (EXPLAINED) SUM OF SQUARES =     5503.0181
CORRELATION COEFFICIENT =  0.987
COEFFICIENT OF DETERMINATION =  97.463 PERCENT
```

CHAPTER 2

Computers and Computing

2.1 Introduction

A computer is an electronic device capable of storing numerical data and of performing certain operations on that data. The usual operations are those of ordinary arithmetic, that is, addition, subtraction, multiplication and division. The only other operations the computer can perform, apart from receiving data from or transmitting results to a user, are those of logical comparison, for example, answering the question: 'is number A smaller than (or larger than, or equal to) number B?' Although they are often termed 'electronic brains', computers do not think for themselves. Instead, they follow instructions given to them in the form of a program. Any intelligence they appear to exhibit is a reflection of the intelligence of the person who wrote the program. The fact that a Russian computer 'won' a chess tournament does not mean that Russian computers are cleverer than British or American computers. It simply means that Russian computer programmers are better at programming chess games than their counterparts in other countries.

Programming a computer requires a basic understanding of how the computer carries out its functions of (a) storing numerical data, (b) operating upon this data either arithmetically or by logical comparisons, and (c) receiving data from and transmitting results to a user. This chapter consists of a brief history of the development of computers and a general survey of the way a modern computer performs the functions mentioned above.

2.2 Historical Survey

The idea of a computer is not new. If the computer is thought of as an aid to calculation then the abacus of the ancient Greeks or the mechanical calculators of Pascal and Leibniz could be said to be the forerunners of the modern computer. The concepts on which the modern computer is based are thus almost as old as the idea of counting. Until the nineteenth century, however, calculating devices and machines relied on the human brain to store instructions, data and intermediate results. The calculator itself helped only in the arithmetical process.

An Englishman, Charles Babbage, was the first person to consider the possibility of building a calculating machine which would store the instructions necessary to solve a given problem and also the data required in its solution. Babbage, who was born in Totnes, Devon, in 1791, was an eccentric inventor and mathematician. In the early 1820s, when the Industrial Revolution was beginning, he conceived the idea of using mechanical processes to help in the tedious job of calculating. (It is said that a remark by a colleague, engaged in computing tide tables for the Admiralty, caused him to consider the possibility for the first time.) He began the construction of a mechanical device, which he called the Difference Engine, with financial help from the British government. This engine was to interpolate values in tables (such as tide tables) by the method of differences. By 1833 Babbage had thought of a more grandiose scheme, and the Difference Engine was abandoned. The new scheme involved building a machine that would perform many types of computation, instead of simply interpolating in tables. Babbage called the new machine the Analytical Engine. It had five components: a memory or store, an arithmetic unit or mill, an input section, an output section and a control unit. The store was to be capable of holding 1000 fifty-digit numbers, using a ten-toothed cog to represent the digits 0–9. Fifty cogs were required to hold each number. These numbers were to be transmitted to the mill by mechanical linkages. Once in the mill they would be added to, subtracted from, multiplied or divided by other numbers which had been brought from the store and the result of the calculations was to be transmitted back to the store by mechanical processes. Numbers to be entered into the store would be read from brass plates or cards, and any number held in the store could be printed by the output section. The Analytical Engine had an important feature which other calculating devices lacked. This was the control unit. Instructions were to be read from cards and interpreted by the control unit, which would then set in motion the various levers, cogs and connecting rods required in executing the instructions.

Most of our knowledge of the Analytical Engine comes from the writing of two people: an Italian army officer, Count Menabrea, and Lady Lovelace, the daughter of Lord Byron. Menabrea attended a lecture given by Babbage and took extensive notes. These were organized, translated into English and expanded by Lady Lovelace, who published them in book form in 1842. Her book has been described as the first programming manual to be written. Although the Analytical Engine, like the Difference Engine, was never finished, the ideas that lay behind it were of great importance over a century later when the availability of electric power and advances in engineering made it possible to build the modern equivalent of Babbage's Analytical Engine. An excellent account of Babbage's life can be found in P. and E. Morrison, *Charles Babbage and His Calculating Engine*, Dover Books, New York, 1961.

The military requirements of the Second World War provided the stimulus for considerable research into the problems associated with building electronic computers; one of the earliest of these was used for several years by the U.S. Army to compute ballistics tables. British universities, notably Manchester and Cambridge, were in the forefront of computer design and construction at this time. The first commercial computer in the United Kingdom was completed in 1951 by Ferranti Ltd. in association with Manchester University. The UNIVAC 1 and the IBM 650 computers were available in the United States a year earlier, in 1950. Since that time, computers have become more powerful, more expensive (and more widely misunderstood) with every year that has passed. For example, the computer complex installed in 1972 at the Manchester University Regional Computing Centre cost £3.25 million, and is about 20 times more powerful than the ICL Atlas computer that it replaced. At the other end of the scale, modern 'minicomputers', costing about £10,000, have greater capabilities in terms of store and speed of execution of instructions than the biggest digital computers of the early 1950s. All modern computers still incorporate the basic five components of the Analytical Engine. These components will be discussed in Section 2.4.

2.3 The function of binary numbers

Before examining the components of the modern computer it is necessary to digress somewhat and consider the way in which numbers are held inside the store of a modern computer. In Babbage's Analytical Engine, a ten-toothed cogwheel was to be used to hold one of the digits 0–9. Fifty such cogwheels in a group could hold a fifty-digit number, in the same way as a set of wheels represents the digits on a car's mileometer. Electronic computers use the properties of electro-magnets rather than mechanical devices to hold numbers. An electro-magnet has two poles, north and south, and the direction of polarity can be determined by sensing wires. The direction of polarity can be reversed, and so the magnet can be in one of two states, compared to the ten possible states of the ten-toothed cogwheel. Hence it can only hold information corresponding to two rather than ten digits. The digits 2, 3, 4, ..., 9 cannot be represented directly by a magnetic medium. Instead a number system using only 2 digits—the binary number system—is used to represent numbers. It employs only the digits 0 and 1.

Before proceeding further with a description of binary numbers, the idea of a number system will be considered using the familiar decimal (base ten) system as an example. The number 1758 can be broken down into four components: eight units, five tens, seven hundreds and one thousand. This could be represented by the following table:

10^5	10^4	10^3	10^2	10^1	10^0
0	0	1	7	5	8

The symbol 10^n means: ten multiplied by itself n times. Thus, $10^3 = 10 \times 10 \times 10 = 1000$. The value of 10^0 is 1. The number 10 is raised to successive powers (0, 1, 2, ..., n) because 10 is the base of the decimal number system. There is no reason why any other positive whole number should not be used as the base of a number system. For example, the number 2 could be selected as a base, and a binary number written as:

2^5	2^4	2^3	2^2	2^1	2^0
0	1	0	1	1	0

This is interpreted at $1 \times (2 \times 2 \times 2 \times 2) + 0 \times (2 \times 2 \times 2) + 1 \times (2 \times 2) + 1 \times (2) + 0 \times (1)$ in the same way that the number 1758 in decimal was interpreted as $1 \times (10 \times 10 \times 10) + 7 \times (10 \times 10) + 5 \times (10) + 8 \times (1)$. To show that a number is a binary number it is written with a subscript 2, for example, 10110_2. Decimal numbers are written with a subscript 10, for example, 10110_{10}. Where there is no possibility of confusion these subscripts are usually omitted.

Counting in binary is straightforward if the layout described in the previous paragraph is adopted. The following tables gives binary-decimal equivalents for $1_{10} - 20_{10}$.

Table 2.1 Binary–Decimal equivalents

Decimal	Binary	Decimal	Binary
1	1	11	1011
2	10	12	1100
3	11	13	1101
4	100	14	1110
5	101	15	1111
6	110	16	10000
7	111	17	10001
8	1000	18	10010
9	1001	19	10011
10	1010	20	10100

In the next section the components of the electronic digital computer will be described, beginning with the store and the way binary numbers are used to hold information in the store.

2.4 Components of an Electronic Computer

Babbage's Analytical Engine had five components: the store, the mill, and the input, output and control sections. A modern electronic computer has the same five components, which will now be described in turn.

(1) The Store

The store of an electronic computer consists of a large number of ferro-magnetic cores (like miniature cotton-reels) strung on a network of very fine wires. Each core can be polarised in one of two directions (i.e. N–S or S–N) and this property can be used to hold binary information, as described in Section 2.3. Each core can hold one binary digit (0 or 1) and a set of several cores is used to hold a binary number. For example, four cores would be needed to hold the binary numbers 0000 to 1111 (decimal 0 to 15) (see Table 2.1). A set of cores is termed a computer *word*, a *store location* or simply a *location*. The length of the word is expressed as the number of *bi*nary digi*ts* (bits) that can be held in that word. For example, one could speak of a '36-bit word'. Most computers have a fixed word length of 24, 32, 36, 48 or 60 bits depending on the manufacturer. For example, an IBM 360 computer has 32 bit words and an ICL 1906A computer has 24 bit words. The store of a typical modern computer holds 32K to 256K words. The symbol 'K' means 2^{10} or 1024, so 32K is equal to 32,768.

Each word or group of cores within the memory can hold a binary number. Whole numbers (*integers*) can be converted directly from base 10 to base 2. For example $16_{10} = 10000_2$. Usually the leftmost bit of each word is used to indicate the sign of the number held by that word. If the word was 16 bits long then $+16_{10} = 10000_2$ would be held as

bit number	16	15	14	13	12	11	10	9	8	7
contents	0	0	0	0	0	0	0	0	0	0

bit number	6	5	4	3	2	1
contents	0	1	0	0	0	0

the '0' in bit 16 indicating that the number is positive. A '1' in this position would indicate a negative number.

Real numbers expressed in base 10 cannot be represented directly in the computer in terms of 1s and 0s, so a method of conversion is employed. A real number can be broken down into two parts, termed the *mantissa* and the *exponent*. Thus, the base 10 real number 2.76 could be broken down into a mantissa (.276) and an exponent (1), meaning:

$$2.76 \equiv .276 \times 10^1$$

The base 10 is raised to the power of the exponent 1 because the number is expressed in the base 10 number system. Notice that if the decimal point is placed before the first nonzero digit of the mantissa, only two numbers need to be remembered, and both are whole numbers. If the base of the number system is 10, then the number 2.76 can be reconstructed from the two whole numbers 276 and 1 if it is remembered that the decimal point comes before the first nonzero digit of the mantissa (276) and that the base of the number system (10) is raised to the power of the exponent (1), i.e. $.276 \times 10^1 = 2.76$. Thus, a base 10 real number can be represented by two base 10 whole numbers. The same is true of base 2 real numbers. A base 2 real number can be converted into two base 2 whole numbers, each of which can be directly represented in the store in the form of a base two whole number. The two whole numbers (mantissa and exponent) are held in one computer word, and the control unit will recover the sign, the mantissa and the exponent when the contents of that word are accessed. Thus, if an integer location is accessed, then a direct base 2 to base 10 conversion is carried out. If a floating point location is accessed, then the pattern of 1s and 0s is split into the mantissa and the exponent, both of which are converted into a base 10 real number. One source of error is to mix the two types of storage (floating point for real numbers and integer for whole numbers). Although logical in ordinary arithmetic, the expression $6.3 \div 4$ will not necessarily produce the right answer if used in a computer program, as 6.3 is a real number and

4 is a whole number. The correct way to program this expression is to write 6.3 ÷ 4.0.

The distinction between integer and real numbers is strictly adhered to in computer programming. For example, 4.0 is stored as a floating point quantity because it has a decimal point, even though it is quantitatively equal to 4, which is stored as an integer number. In computer programming terms, whole numbers are called integers and other numbers (even of the form 4.0 or − 16.0) are called floating point numbers (because of the way the decimal or binary point is handled, as explained earlier). Both are stored as patterns of binary digits within the computer word, and the computer cannot distinguish between the two unless the programmer instructs it to do so. The pattern of bits for an integer number is (sign bit) (binary number) whereas for a floating point number it is (sign bit) (exponent) (mantissa).

The store can hold alphabetic as well as numerical data. This is done by assigning numeric codes to the letters of the alphabet. For example, A might be represented as 32_{10} ($=010000_2$) and B as 33_{10} ($=010001_2$). Six bits are usually sufficient to provide enough codes for the 26 letters of the alphabet plus the characters ( ) + − * £ . , etc. Depending on the word length of the computer, a maximum of four or more characters can be held in each word (though less than four characters can be stored per word if required, the unused part of the word being set to zero). As with integer and floating point numbers, the codes representing characters appear to the computer simply as binary patterns. The programmer must instruct the computer to treat the contents of a particular word as an integer number, a floating point number or a set of characters.

The store of an electronic computer is thus seen to consist of a large number of ferrite cores which are grouped together to form words. Each word can hold an integer number, a floating point number or a maximum of four or more characters. A typical modern computer has 32K to 256K words of store, and the contents of any word can be accessed in a fraction of a second (usually millionths or tenths of millionths of a second). This is termed the access time.

For some applications, for example in survey analysis or in automated cartography, large amounts of data are handled. The size of the store may then be inadequate. The cost of extra core store is relatively high; also, the size of store that a given control unit (see Section 2.4.(4)) can handle is limited. A cheaper form of mass storage involves the use of magnetic tape, disc or drum. The penalty is slower access time, which may be ten or more times greater than the access time for the central store. This additional store is thus called slow store or backing store. It can be used not only to hold extra information required temporarily by one particular program, but to hold information permanently as, for example, data sets or program libraries. The backing store can be broken down into words in the same way as the store. A group of words holding one set of information is termed a file. A program library might form one file, and the census data for a particular region another.

A distinction is made between files that can be accessed or searched randomly and those which can be accessed serially. A random-access file is stored on a disc or a drum, and any word within the file can be found more or less as rapidly as any other. A serial file usually exists on magnetic tape but can be simulated on discs or drums. As the name 'serial' implies, information is stored (and retrieved) sequentially so that, to reach the 1000th word of a file the previous 999 words must be read, presuming the reading head is located initially at word 0. Finding a particular track on a gramophone record is analogous to locating a word in a random access file, for the movement of the turntable and the movement of the user's hand can, in theory, allow any point on the record to be located in approximately the same time. If the same record were taped, then finding the same track on the tape would involve winding and rewinding the tape.

Some computers, such as the IBM 370 range, use a memory system in which the central store and backing store are closely integrated so that the size limitation of the central store need no longer be a problem. These computers are said to have a virtual memory system. In fact, a British computer, the ICL Atlas, was the first to use this method in the early 1960s.

(2) Input and Output Devices

Communication between the user and the computer is dependent upon the ability to input or output information. The most common input unit is the *card reader*. As we saw earlier in this chapter, Babbage's Analytical Engine was designed to read information from cards. Card-sorting equipment was also widely used from the latter half of the nineteenth century onwards. Cards are prepared on punches like that shown in Figure 2.1. Each letter, digit or character has an individual card code. For example, 1 is coded as a hole punched in the row 1 of the particular card column. A card has 80 columns, and so can hold a maximum of 80 alphanumeric characters

Figure 2.1

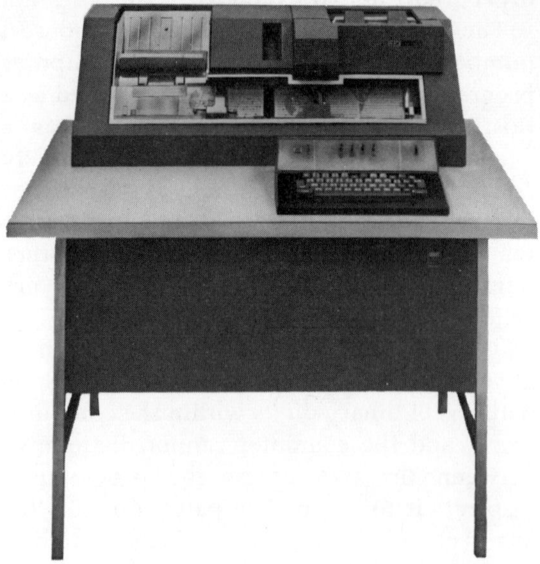

(the 26 upper case letters, the digits 0–9 and special characters such as +, −, ; and ∗). Each column has 12 punch positions numbered (from top to bottom) 12, 11, 0, 1, 2, ..., 8, 9. The row numbers 0–9 are usually printed across the card, and rows 11 and 12 are above the printed section. A punch card holding all 64 characters in ICL 1900 Card Code is shown in Figure 2.2. Other computer manufacturers

Figure 2.2

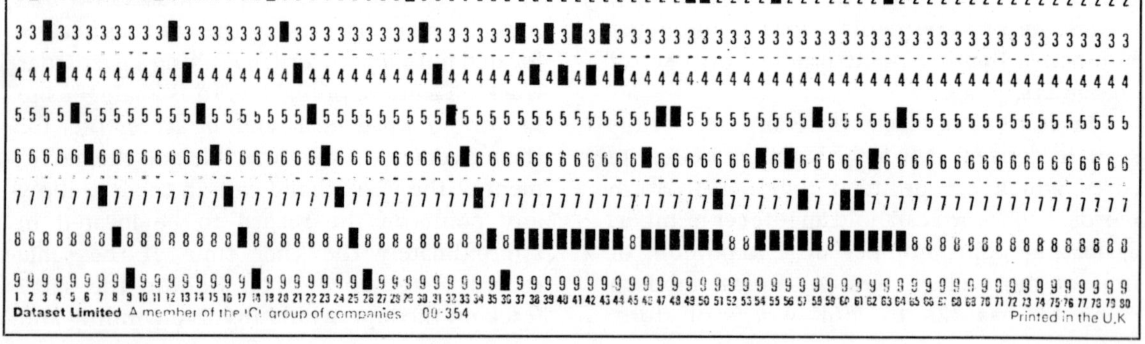

have their own card codes, so that a deck of cards prepared for one computer may be unusable on another machine of a different make. The cards are read by a card reader, such as that shown in Figure 2.3, by means of

Figure 2.3

a photoelectric cell. A light is shone through the reading station and if there is a hole in the card this light reaches a photoelectric cell opposite the hole. Thus, the pattern of holes on the card is converted to an electrical signal which is transmitted to the core store as a card image. The process of reading cards can be subject to error if the cards are bent, warped or worn so, if the card deck is sizable, it is good practice to keep a duplicate set. A less frequently used card is the mark-sense card which is not punched but which is marked in pencil in the appropriate positions. Such cards can be used in field surveys and have the advantage that records can be transcribed to punch cards by machine. This is often faster than the coding of questionnaires and manual punching of the information recorded on the coding sheets.

Other input devices include the *paper tape reader,* which decodes sequences of holes punched across a narrow ribbon of paper tape much in the same way as the card reader decodes sequences of holes punched across a card (Figure 2.4). Most users find that paper

Figure 2.4

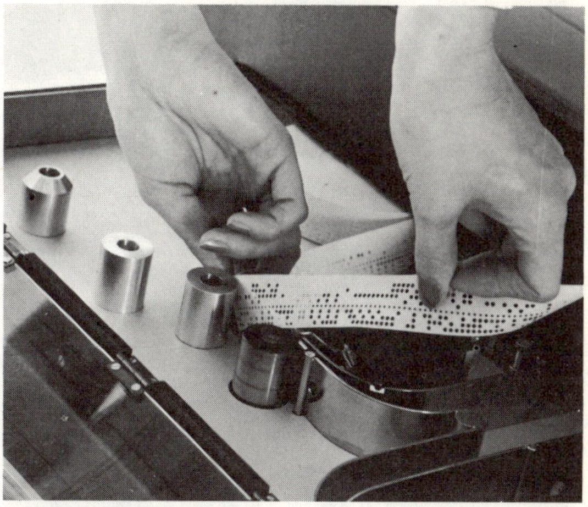

tape is more suited to transferring information from one computer or data generating or collecting device to another computer than to acting as a general input medium. For example, automatic river-level recorders punch the water-level measurement every 15 minutes or so on to a roll of paper tape which is replaced every few days. This tape is easy to carry and relatively cheap. However, correcting one line of a program or one data value where both are punched on tape is much less easy than altering one card.

The *teletype* is an electric typewriter which is connected directly to the computer, thus allowing the user to transmit information directly from the keyboard without the need for some intermediate form of storage being used. Some teletypes are usable in off-line mode (disconnected) as paper-tape punches but their main usage is in on-line mode (actively connected to the computer); in on-line mode they are used to edit (amend, or modify) files which are stored in the fast or (more usually) the backing store of the computer. The cost of such usage is high, however, and is normally restricted to experienced users.

Magnetic tapes and discs can also be used

as input media. A tape unit is shown in Figure 2.5. Small cassette tapes, similar to those used

Figure 2.5

in commercial tape-recorders, can be attached or *interfaced* to automatic data-collection or remote-sensing equipment, and the resulting records can be read directly by the computer. The standard tapes and discs, which can form the computer's backing store, can be used to transfer large amounts of data from one computer to another. A large program library can be stored on one standard magnetic tape which can be sent by post or otherwise from one computer installation to another.

Output devices include the *lineprinter* and the *graphplotter,* though devices such as teletypes, card punches, paper tape punches or magnetic tape decks can be used for output. A relatively recent innovation is the CRT (*Cathode Ray Tube*) or VDU (*Visual Display Unit*) (Figure 2.6) which provides a (temporary) display of information output by the computer. Because of its expense, and the fact that output cannot be preserved except by photographic means, VDU's are not widely

Figure 2.6

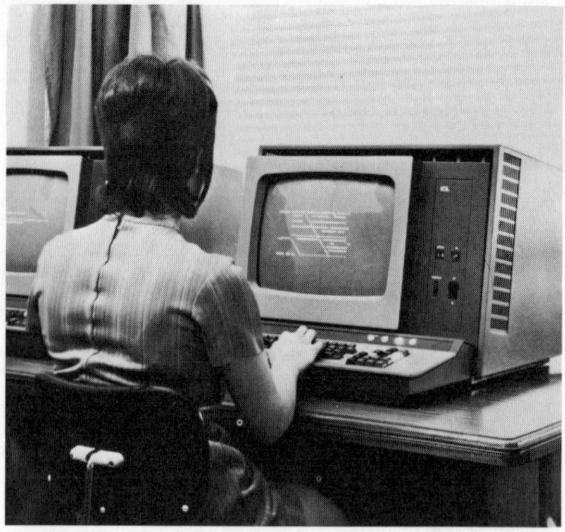

used except by programmers with special requirements, such as on-line editing. The lineprinter (Figure 2.7) is capable of printing

Figure 2.7

output at an impressive rate—200 to 1000 lines per minute is the usual range. Each line, of 120, 132 or 160 characters, is printed not character by character as on a typewriter but as one single string. Pseudo-graphical output, such as scatter diagrams, histograms and isarithmic maps can be printed by the lineprinter. More accurate graphical output is produced by the graph-plotter (Figure 2.8)

Figure 2.8A Drum Plotting System

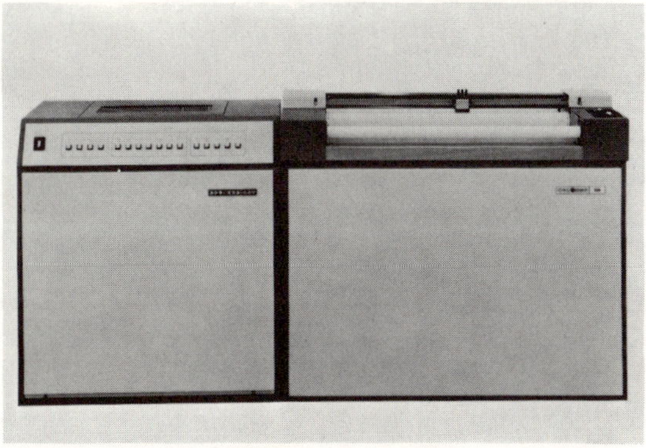

Figure 2.8B Three-pen System

which is capable of drawing contour maps, electrical circuits and plans with high precision. Examples of an isarithmic map produced by the lineprinter and a contour map drawn on a graph-plotter are shown in Figures 2.12 and 1.1.

(3) Arithmetic Unit

Information read into store by one of the input devices is usually operated upon in some manner before being output. In a digital computer these operations are either arithmetic (e.g. addition) or logical (e.g. comparison of the magnitude of two numbers) though both kinds of operation are essentially arithmetic. Calculations and logical comparisons are performed electronically, often by specially wired circuits, and intermediate and final results are stored in an accumulator. Medium and large computers may have several accumulators, allowing simultaneous calculations to be performed. The results of the calculations are returned from the accumulator to the store. For example, the instruction 'add the contents of word number 1 to the contents of word number 2 and put the result into word number 3' involves the transfer of the contents of word number 1 to the accumulator, then the addition of the contents of word number 2 to the contents of the accumulator and finally the transfer of the new contents of the accumulator to word number 3. The original contents of word number 3 are lost in this transfer; they are said to be *overwritten*.

(4) The Control Unit

The functioning of the input-output devices, the backing store and the arithmetic unit depend upon the existence of a supervisory or control unit, which co-ordinates machine activity and regulates transfers into and out of the memory. The control unit and the arithmetic unit are together known as the *central processing unit* (CPU), central processor or mainframe while input-output devices, such as card-readers, lineprinters, disc units and magnetic tape units are termed *peripherals* or peripheral devices. A typical computer installation is shown in Figure 2.9 (see p. 14). The CPU is the nerve-centre of the computing system. In a large modern computer, which may run several programs simultaneously, the control functions are extremely complex.

Interested readers should consult Hollingdale and Toothill (1965) for a more expanded yet readable account of the material presented in this chapter.

Figure 2.9

2.5 The Functioning of the Computer

In the preceding section the component parts of the computer were described. In this section the way the computer works as an integrated whole is summarized. It will be assumed that the computer described in this section has a large core memory (store) of 32K words or more, with backing store (magnetic tapes, discs or drums) and accepts input from a card reader and sends output to a lineprinter. This system is shown diagrammatically in Figure 2.10 (see p. 15).

Each separate job that is fed into the computer consists of three sections:

(i) Job Control Section
(ii) Program
(iii) Data

Some programs do not require data, for example, simulation programs, but most of the programs in this book will read data from the card reader.

The Job Control Section contains information which specifies the name or identity of the user and the facilities required by the program. These instructions are given in a language called Job Control Language (JCL). Some JCL's are very complex (for example the GEORGE JCL for ICL 1900 series computers) but the JCL needed for the programs given in this book will be simple and will be fully described in the documentation produced by

Figure 2.10

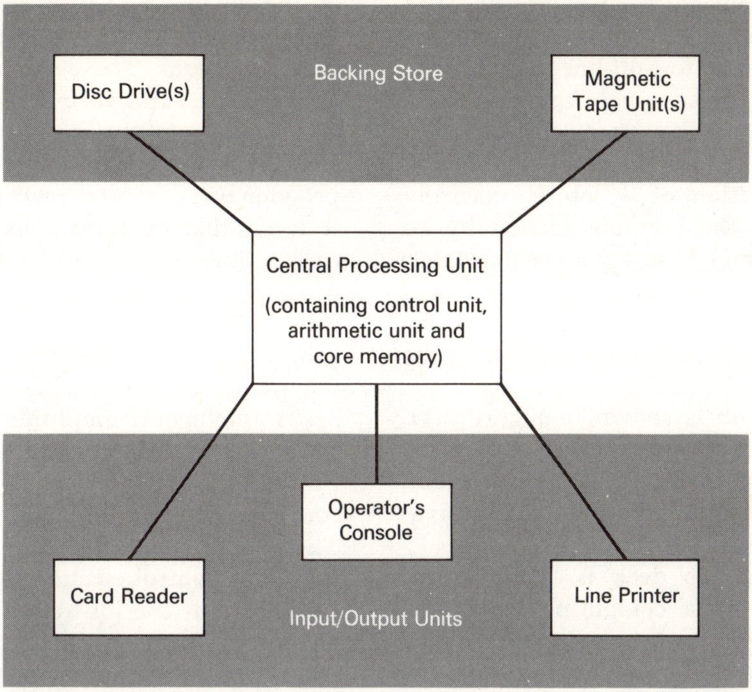

your local computing centre.[1] For example, the JCL required to run a job on a DEC PDP-11 minicomputer is:

```
$JOB CALCUL [222,222]
$EXECUTE
```

These cards indicate that a new job is to be commenced, that its name is CALCUL (any 6 letter name could have been used) and that the user has the code [222,222]. $EXECUTE instructs the computer to run the job.

The second section of the job is the program. A program is a set of instructions that tell the computer how a given problem is to be solved. For example, a program to add two numbers stored in words 1 and 2 has already been given. Most programs are written either in a language called Fortran (*For*mula *Tran*slator) or in a language called Algol (*Algo*rithmic *L*anguage). These are languages that appear to be similar to ordinary algebraic formulas. For instance

```
A = B + C
```

and

```
Z = 10.5 + D - X
```

are Fortran instructions. Algol instructions look more like English:

a: = b + c; **for** i = 1 **step** 1 **until** n **do** . . .
z: = 10.5 + d − x; **if** (x) **then** . . . . .

if *(x = y)* **then** *z: = 10*

and

a: = b + c

Other programming languages have been invented. Some are for special purposes, for example Cobol is a commercially-orientated language and Simscript is a language designed for use in simulation programs. Others are

[1] You should consult the computing centre documentation at this stage to find out the appropriate JCL for simple Fortran programs.

scientific languages (like Fortran and Algol) but are less popular. Examples are PL1 and APL. Some computers can use a language specially designed for on-line use from a teletype. The best known of these is Basic, but others are widely used, for instance, Focal.

The data section of a job is normally separated from the program section by an indicator **** and # # # # S are examples) or else by a JCL instruction, such as $EOD or DATA. The data cards contain the information that is required in determining the particular results required.

A complete job is shown in Figure 2.11. The complete set of cards is called a job deck. If the contents of each card were written out, at one line per card, they might appear as shown in Table 2.2 (see p. 17).

The complete job deck is taken to the reception area of the computing centre and handed to an operator, who will note the time the job was handed in, the job name and the priority of the job. Next, he will load the job deck into the card reader. The control system of the computer will then transmit instructions via the control unit to cause the job deck to be read into the store. The control system is really an elaborate program that carries out some of the functions of the computer operator. With early computers the operator had to instruct the control unit directly by setting switches on the front panel of the computer. Later, complex programs were developed that would perform many of these functions. These programs are known as executive, supervisor or monitor programs, and they form the basis of the operating system. In a modern computer the monitor program will control the work of the machine. It is not like a human operator in that it carries out automatically a pre-set sequence of instructions. Sometimes these instructions are so numerous that they can give the monitor program the appearance of being intelligent. Computers are not intelligent, though; they act as 'intelligence amplifiers'.

The job deck has, so far, been transferred by the monitor from the card reader to the core memory in the form of electrical impulses. The next stage involves the interpretation of the Job Control section to find out the facilities that the job requires, for instance, an output device (usually the lineprinter), and access to the library of instructions that the monitor controls.

One of the facilities that the job described above would request would be the Fortran *compiler*, because the program instructions are written in Fortran. The Fortran compiler is another program, controlled by the monitor, that interprets the user's program and converts it into instructions that the control unit can execute. The Fortran instruction

Figure 2.11

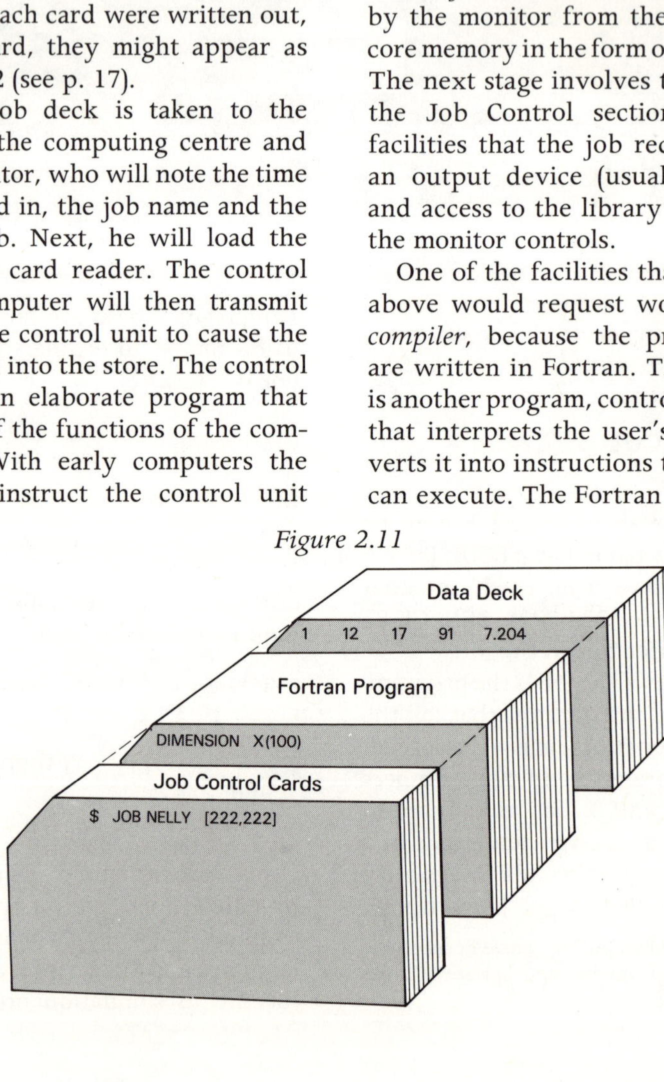

Table 2.2

Job control

```
$JOB DIST[222,222]
$ASSIGN BI:,5
$ASSIGN LP:,6
$RUN FORTRN
#OBJECT,LP:<BI:
```

Program

```
      DIMENSION X(10)
      READ(5,1) N
    1 FORMAT(I3)
      READ(5,2) (X(I),I=1,N)
    2 FORMAT(F10.0)
      S=0.0
      DO 3 I=1,N
    3 S=S+X(I)
      WRITE(6,12) S
   12 FORMAT(8H0SUM IS ,F10.4)
      STOP
      END
```

Job control

```
$RUN LINK
#LOADER<OBJECT,FTNLIB/L/E
$RUN LOADER
```

Data

```
  5
1.0
2.0
3.0
4.0
5.0
```

Job control

```
$FINISH
```

$$Z = A + B$$

is interpreted as:

(i) set the contents of an accumulator to zero;

(ii) bring the contents of the word called A to the accumulator and add them to the current contents of the accumulator;

(iii) repeat step (ii) using the word called B instead of the word called A;

(iv) put the current contents of the accumulator into the word called Z.

Each of these four components is a machine instruction, and is held in the store in the form of a binary number (Section 2.3). Each Fortran instruction is decoded in this way, and the result is a compiled program in binary form. Some Fortran instructions may involve an operation such as finding the square root of a number, or calculating the logarithm of a number. These operations are used in many programs, and ready-compiled instructions for each of these commonly-used operations are kept in the system library. The second stage of preparation involves searching this library for the required instructions. This stage is known as linking or consolidation.

The linked binary program is a complete set of machine-code or binary instructions, each of which is capable of being executed by the control unit. Once it has been linked, the program is executed or run. During the execution phase the data is read from the card reader and output is sent to the line-printer. On termination, the binary program can be preserved as a file on backing store or else deleted.

This description of the working of a computer is far from complete. Modern large computers can compile, link and run several programs at once, as well as support on-line editing from teletypes or VDUs located in many parts of the university or college. Telephone links can be used to allow remote users to transmit jobs from teletypes or from remote job-entry terminals (which consist of a small control unit, a card reader and a line printer). In the United Kingdom, regional computing centres (such as the Manchester University Regional Computing Centre) can handle jobs which are read in by the card readers of remote job entry terminals at other universities and colleges which are up to 150 kilometres away. Some military and commercial users transmit jobs by satellite link from Europe to computers in the United States. Computer networks, which may involve several computers, are being developed rapidly at the moment, partly to reduce costs by siting one large computer centrally and linking other smaller ones to it, and partly to provide cover in case one computer in the network breaks down.

Another feature of computers that was not mentioned in the description above is the way in which input can be read not only from a card reader but also from paper tape, magnetic tapes, teletypes, or cassette tapes. Output can be directed to files in the backing-store, to card punches, to the lineprinter, or

Figure 2.12

KEY TO CONTOUR VALUES

......	-175.0000	TO	-100.0000
444444	-100.0000	TO	-25.0000
......	-25.0000	TO	50.0000
555555	50.0000	TO	125.0000
......	125.0000	TO	200.0000
$$$$$$	200.0000	TO	275.0000
......	275.0000	TO	350.0000
666666	350.0000	TO	425.0000
......	425.0000	TO	500.0000

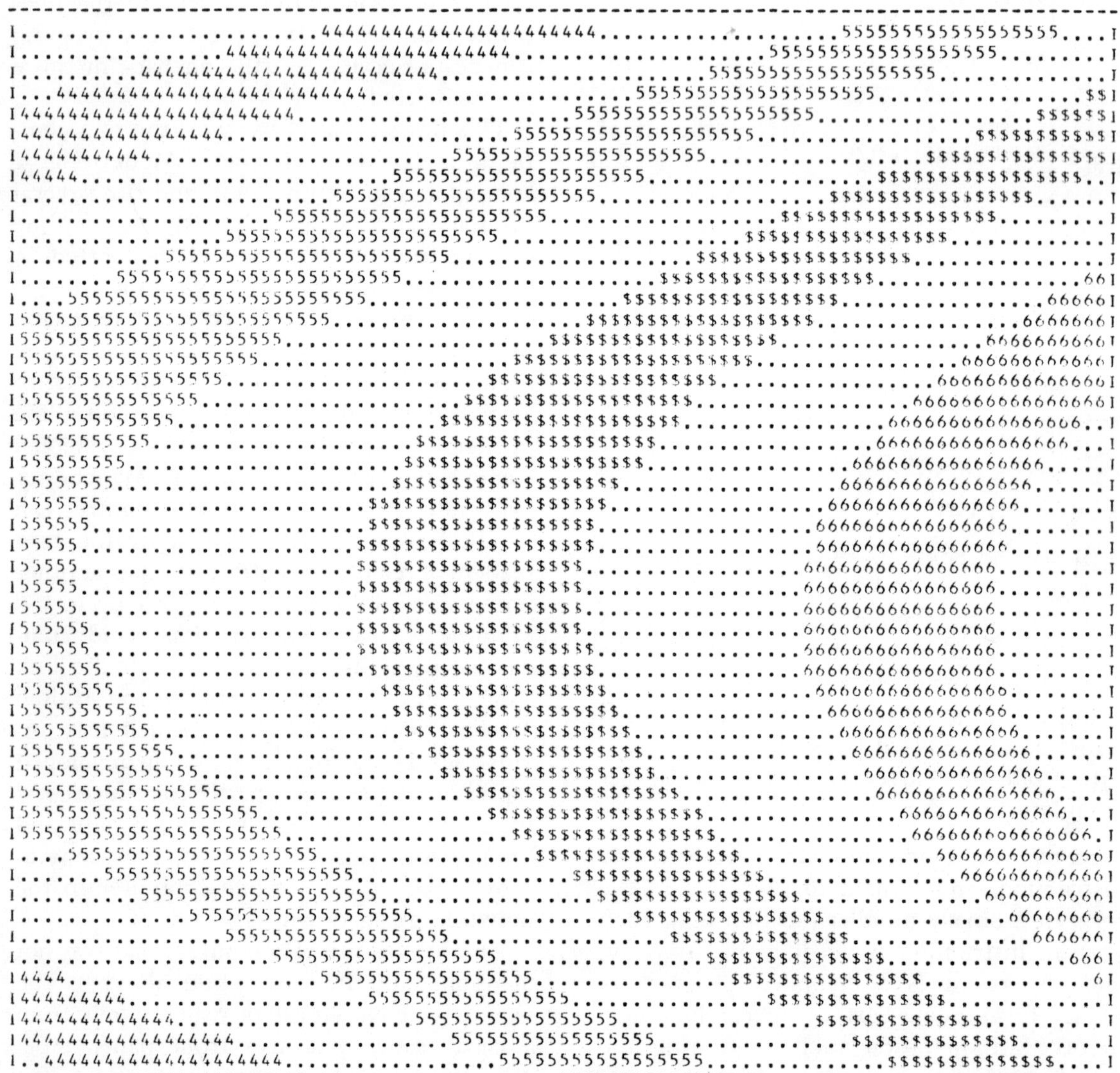

to a graph-plotter. Geographical users often make considerable use of graph-plotters in the preparation of contour-type maps and diagrams (Figures 1.1 and 1.2), though isarithmic maps, histograms and scatter diagrams can be produced by the lineprinter (Figure 2.12).

In the following chapters the Fortran language will be described. Before reading Chapter 3, however, you should turn to Appendix A where the operation of a card punch is described. Follow the instructions given in Appendix A and practice for at least ten minutes by punching several cards with, for example, your name or the football results in last Sunday's newspaper. In Appendix A you will also find the layout of a Fortran program card described. Using this description as a reference, punch out the program given in Table 2.2 above, together with the appropriate job control instructions and data cards (the layout of which is also given in Table 2.2). As a test of your copying ability, run the job and see what the computer makes of it. Remember to use the JCL for your own computer—the JCL in Table 2.2 is appropriate for PDP-11 computers only.

CHAPTER 3

The Fortran Language

3.1 Introduction

Fortran is one of the many 'high level' computer programming languages. It is preferable to others, because (i) some languages are highly 'machine dependent', that is, they will work only on one particular machine or range of machines (PL1, APL and Atlas Autocode are examples), and (ii) many programs that are widely available to geographers are written in Fortran (see Appendix B for a listing of sources of computer programs). Fortran, in its standard form, is not machine-dependent and is defined by a rigorous set of rules (*Standard Fortran Programming Manual*, National Computing Centre, Manchester, Second Edition, 1972). Thus, Fortran programs can be run on all the large computers available in British and American universities, and on many 'minicomputers' such as DEC PDP 11's. You will find that some computer manufacturers have extended their own versions of Fortran to include nonstandard 'extras', but it is safe to say that a standard Fortran program will run on any machine supporting Fortran. (A list of nonstandard facilities is usually included in the manufacturer's own Fortran manual.)

Other differences between the way in which the various makers of computer handle Fortran programs will also be described in documentation produced by your computing centre. For example, ICL 1900 series computers require that the first card of a Fortran program be of the form 'MASTER xyz' where xyz is a name such as PROGRAMTOMAPSPATIALDISTRIBUTIONS. The last card of a Fortran program in ICL Fortran is FINISH. CDC 6600 and 7600 computers also need an introductory card of the form PROGRAM xyz (INPUT, OUTPUT, TAPE5 = INPUT, TAPE6 = OUTPUT) where xyz is a name of up to 6 letters. Again, it will be assumed that you have found out these requirements from the staff of your computing centre or from your course instructor.

Now we can turn to the Fortran language itself. Like any language, there are two ways of learning it. One is to learn from experience. The other is to learn the grammar and vocabulary. This is similar to an English-speaking student attempting to learn German either by going to a German-speaking country for a few months (during which time he would have to pick up the rudiments of the language to survive!) or by reading a textbook. He would eventually learn German by either process, but the method of learning by experience would possibly be more interesting. That is why this book will try to get you to learn Fortran by actually running programs right from the start, rather than by giving you pages of rules and regulations to memorise. This may mean that at some stages you are not sure of the meaning of the instructions you actually punch, but it is hoped that all will eventually become clear. Also, some of the example programs and exercises may seem trivial at first; this is so that you can concentrate on learning the language without worrying too much about equations and formulae, or even what the problem means.

The rest of this chapter is set out in the form of sections. Each section should take you two to three hours to read and understand. Examples and problems are included in each section. You should first of all punch out the example programs, making a note of the meaning of each instruction, and then run them. (Refer first to Appendix A for a description of the way in which program cards and data cards are punched. The use of the card punch is described in Appendix A.) When you have got them to work correctly, try to program one of the exercise problems. In the first few sections, these problems involve only slight modifications to one or other of the example programs, but for later sections you will need to refer to a statistics book, such as the companion to this volume (Ebdon, 1976). Remember that you will never learn Fortran simply by reading a book; you must persist in getting example programs to work. (If nothing else, learning Fortran will test your patience.)

3.2 Input and Output

The most fundamental part of any program involves input and output. It is of very little value to write a program which does not involve the printing out or display of the results. Usually, too, most programs need to instruct the computer to read in data.

The READ and WRITE instructions perform the functions of input and output. Each of these instructions needs parameters (descriptors) to tell it what to read, or write, and in what manner.

In Chapter 2 we saw that a variety of input and output devices were available. Each may have an actual number stamped on it (e.g. 'ICL 2973') but within a program each device can be given a *logical number* by the programmer. By logical number is meant one that is used by the programmer for his or her convenience. For example, the number 5 is usually used in READ statements to specify the card reader and the number 6 is used in WRITE statements to specify the lineprinter. Elsewhere in the program such numbers have no special significance. The operating system keeps a record of all logical numbers and corresponding input/output devices used in a program. In most cases, logical numbers 5 and 6 are assigned to the card reader and lineprinter respectively by default (i.e. the programmer does not have to specify them in the job control). In the rest of this book these default settings will be used.

The second parameter tells the operating system how to locate the position on the card of the number to be read or the position on the page on which a number is to be printed. This second parameter is the label number of a *format statement*. All format statements must have an identifying label, which is any positive integer value less than or equal to 99999[1]. The label is punched in columns 1 to 5 inclusive of the punch card (see Appendix A, Figure A.3). If the label has less than 5 digits, spaces should be put on the left hand side. Thus the label 1 is punched as ΔΔΔΔ1 where Δ means: punch a space. The second part of a format statement is of the form

FORMAT (format specification)

This part of the statement is punched between columns 7 and 72 of the card. Thus, the word FORMAT could start anywhere after column 6 provided that the statement did not continue after column 72. Normally the word FORMAT is started in column 7. Format specifications, which follow the word FORMAT in the format statement, are slightly different for input and output formats (i.e. formats specifying how data is to be read or how results are to be printed, respectively). Both input and output formats have several common characteristics. The first of these is the form

[1]Less than this on some computers.

of the statement, which is

label FORMAT (format specification)

Within the format specification special codes are used to differentiate between floating point and integer quantities (see Section 2.4.1 for a definition of these terms). These codes are F (for floating point) and I (for integer). Each number to be read or printed occupies a certain number of card columns (in the case of input) or lineprinter character positions (in the case of output). Also, one or more numbers may be read from one card or printed on one line. The format specification involves the description of the type of number to be read or printed (floating point or integer), the number of character positions or card columns occupied by each number and the number of items to be read from one card or printed on one line.

F format descriptors have the form nFa.b where F indicates that the number(s) to be read or printed is represented in the computer word in floating point layout, n is the number of items per card or per line, a is the total number of columns or character positions to be allocated to each number and b is the number of columns or character positions that follow the decimal point. If n equals 1 it is omitted.

EXAMPLES

(i) F10.4 means: one floating point quantity to be input or output. It occupies 10 columns or character positions and has 4 decimal places. A data card punched in this format might contain ΔΔΔ76.1397 and a line of print output in this format might contain Δ9234.7200.

(ii) 2F4.2 means: two floating point quantities to be read from a data card or printed by the lineprinter. Each floating point number occupies 4 card columns or character positions and has two places of decimals. Examples of data cards punched in this format are 1.629.35 (read as 1.62 and 9.35) and 0.001.00 (read as 0.00 and 1.00). Examples of lines of print output in this format are 6.723.72 (read as 6.72 and 3.72); 0.900.17 (interpreted as 0.90 and 0.17).

I format descriptors have the form nIa where I specifies that the number(s) to be read or printed are of type integer, n is the number of integers to be read from the data card or printed on one line, and a is the number of columns or character positions each integer occupies. If n equals 1 it can be omitted.

EXAMPLES

(i) I3 means: one integer to be read from a data card or printed on one line. The integer occupies 3 columns or character positions, for example: ΔΔ6 is read as 6, −10 as −10.

(ii) 2I6 means: two integers, each occupying 6 columns or character positions to be read or printed. The numbers 17 and 8 would be punched on a data card as ΔΔΔΔ17ΔΔΔΔΔ8 using this format specification.

Output and input formats, and their use, will now be described separately.

Output statements involve (a) a WRITE statement and (b) a FORMAT statement. The WRITE statement has the form

WRITE (6, label number) names of store ↔
↔ locations holding items to be printed

EXAMPLES

(i) WRITE (6,1) A,B,C. This instructs the computer to output the contents of the computer words identified as A, B and C using the format labelled 1. The '6' in the brackets is the logical number of the lineprinter.

(ii) WRITE (6,10) X,I,J. This instructs the computer to output the contents of the computer words identified as X, I and J using the format labelled 10.

The format specification associated with the WRITE statement must always begin with the four characters *1HΔ*, and it must specify the type of number to be printed (code F or I), the number of such numbers and the size of each number in terms of lineprinter character positions, as explained above.

EXAMPLES *(continued)*

(iii) The format labelled 1 in Example (i) above might be

1 FORMAT (1HΔ, 3F10.4)

if it is assumed that A, B and C are locations containing floating point quantities. For example, if A = 10.4, B = −176.39 and C = 0.7171 the line of output produced by the two statements

WRITE (6,1) A,B,C
1 FORMAT (1HΔ,3F10.4)

would be

10.4000 − 176.3900 0.7171

Notice that zeros after the decimal point are printed in full (e.g. 10.4000 because 4 places of decimals were specified) but nonsignificant zeros before the decimal point are not printed unless the number to be printed is less than 1 (e.g. 0.7071). Also note that the + sign is never printed but the − sign always is, and must be counted as equivalent to one digit in the format specification.

(iv) The format labelled 10 in Example (ii) might be

10 FORMAT (1HΔ,F10.2,2I7)

assuming that X is the identifier of a floating point location and I and J are identifiers of integer locations. Unless otherwise stated, identifiers beginning with the letters I, J, K, L, M and N will refer to integer locations. Identifiers beginning with the letters A–H and O–Z will be the names of floating point locations.

Example Program 1

This program will add together the two floating point quantities 6.3 and 3.7 and store the answer in a location that will be called A, though any identifier could in fact be used, subject to limitations that will be discussed later. The contents of location A will then be printed out. The printout produced by this program consists of one number, 10.00, which is printed in the first five character positions of a line. (Note: the decimal point counts as one character.)

Table 3.1

```
   A = 6.3 + 3.7
   WRITE (6,10) A
10 FORMAT(1H ,F6.3)
   STOP
   END
```

EXERCISES

(1) Modify Example Program 1 to cause it to instruct the computer to add together 17.2 and 97.63. You will have to change the format specification slightly so that the answer (114.83) can be printed. This number requires a total of 6 character positions, and has two decimal places.

(2) The statement A = 6.3 + 3.7 in Example Program 1 instructs the computer to add 6.3 and 3.7 and to place the result in a store location called A, replacing the previous contents of location A. The symbol '—' is the subtraction instruction, as in arithmetic and algebra. Write a short program to subtract 10.1 from 19.4 and print the result using an appropriate format. Remember that (19.4 − 10.1) will result in a floating point answer, and that identifiers or names of locations holding floating point numbers must start with a letter from the range A–H and O–Z.

The characters *1HΔ*, at the beginning of an output format specification instruct the operating system to begin printing on the next line. Sometimes double-line spacing is required, in which case the first characters of the output format specification are *1H0,*. To begin printing on the first line of a new page the sequence *1H1,* is used. The characters Δ,0 and 1 are called carriage control characters. They control the spacing between lines of printout.

Two or more different format descriptors can be used within a single format specification. For example

```
100   FORMAT (1HΔ,F10.4,I6)
```

specifies that a floating point number with four decimal places will occupy the first 10 columns of the next line, and that columns 11–16 will be occupied by an integer. Printing will commence on the next line because the carriage control character is 1HΔ,. Each format descriptor must be separated from the preceding descriptor by a comma. There is no need to put a comma after the last item in a format specification.

Input statements involve (a) a READ statement and (b) an associated FORMAT statement. The READ statement is of the form:

READ (5, label of format statement) list of ↔
↔ names of locations in which the numbers ↔
↔ are to be stored.

The numbers read from the data card will be stored in the locations identified by the names in the list. For example, the instruction

```
      READ (5,101) A,B,C,D
101   FORMAT (4F10.0)
```

will cause four numbers to be read from the next data card and stored in the locations named A, B, C and D. The FORMAT statement associated with the READ statement does not include initial carriage control characters. In most other respects, the input format and the output format are identical.

EXAMPLES

(i)
```
      READ (5,10) A,J
10    FORMAT (F10.0,I6)
```

These input statements will cause two numbers to be read from the next data card. The first number is a floating point number, the second an integer. (Notice the identifier of the locations in which these numbers are to be stored—A is in the range A–H and J is in the range I–N.) The first number will be punched in the first 10 columns of the data card. If no decimal point is punched it will be assumed to lie after the tenth digit. (Blanks are read as zeros.[1]*) The second number is punched in columns* 11–16 *and will be right-justified, i.e. punched as far to the right as possible. For example, if the two numbers to be read are 6.9423 and 67 the data card could contain one of the following:*

6.9423ΔΔΔΔΔΔΔΔ67
Δ6.9423ΔΔΔΔΔΔΔ67
ΔΔ6.9423ΔΔΔΔΔΔ67
ΔΔΔ6.9423ΔΔΔΔΔ67
ΔΔΔΔ6.9423ΔΔΔΔ67

Optionally, zeros could be punched instead of blanks.

(ii)
```
      READ (5,6)I,J,K,L,M,N
6     FORMAT (6I1)
```

Each of the first six columns of a data card will contain an integer number. The first number will be stored in location I, the second in location J and so on. If the numbers are 6, 7, 1, 0, 5, 0 the card could contain

671050

Alternatively, use could be made of the fact that a blank is read as a zero[1]*, and the card would then contain*

671Δ5Δ

[1] If you use a CDC computer and leave all the card columns of a particular field blank then the number will be read as −0.0 not 0.0 (or −0 instead of 0).

Example Program 2

This program will read five floating point numbers, each having two places of decimals, into locations A1, A2, A3, A4 and A5. The contents of these locations will then be added together and the result placed in location V. The contents of V will then be printed on the first line of a new page, using the carriage control sequence 1H1,.

EXAMPLE

Assume that the following five numbers are to be used as data: 1.62, 19.38, −2.75, 0.75, 1.00. The data card could be punched as follows:

ΔΔ162Δ1938Δ−275ΔΔΔ75ΔΔ100

using the fact that, with format (F5.2) the decimal point will be placed by the operating system immediately prior to the second digit from the end of each set of five digits (columns). Thus, ΔΔ162 is read as 1.62. Alternatively the decimal point could be explicitly punched. For example, the same five numbers could be punched as:

Δ162Δ19.38−2.750.75Δ1.

Table 3.2

```
  READ(5,6) A1,A2,A3,A4,A5
6 FORMAT(5F5.2)
  V=A1+A2+A3+A4+A5
  WRITE(6,1) V
1 FORMAT(1H1,F8.2)
  STOP
  END
```

EXERCISES

(3) Use Example Program 2 to find the sum of six numbers of your choice.

(4) The symbol for multiplication in Fortran is the asterisk * (to avoid confusing the algebraic symbol x and the letter X). Write a program that will instruct the computer to read two integers into locations IA and IB, multiply them together and put their product into the location IPROD. Finally, write instructions to output the contents of location IPROD on the next line of printout. (Hint: the multiplication statement is

IPROD = IA*IB).

(5) Can you find any errors in the following input/output statements? The answer to the first problem is given as a guide.

(a)
```
      WRITE (5,1) A,B
    1 FORMAT (2F10.4)
```

Errors: (i) logical number 5 refers to the card reader not the lineprinter. The statement should be WRITE (6,1).

(ii) There are no carriage control characters in the format specification, which should be (1HΔ, 2F10.4) (or 1H0, or 1H1,).

(b)
```
      READ (5,10) X
   10 FORMAT (I6)
```

(c)
```
      WRITE (6,9) I,J,K,
    9 FORMAT (1H1,,3I10.4)
```

(d)
```
      READ (5,8) IJ
    8 FORMAT (1H0,F10.12)
```

SUMMARY OF INPUT FORMAT INFORMATION

(i) each number to be read from a card is associated with a certain set of card columns;

(ii) if the number to be read is to be stored in a floating point location its format code is F; if an integer, the code is I;

(iii) the decimal point in floating point numbers can be punched or implied; a punched point always over-rides the default (implied) position;

(iv) a number before an I or an F code specifies the number of numbers of that type;

(v) the number following an I code specifies the number of columns each integer occupies.

These columns are treated as units, tens, hundreds, thousands, etc., reading from the right of the specified field. Thus, ΔΔΔ1 is read in (I4) format as 1. In the same format, 1Δ1Δ is read as 1010. In other words, any blank within a specified set of columns is read as zero;

(vi) the number following an F code is of the form a.b (e.g. F10.2). The part before the point, a, specifies the total number of card columns occupied by the floating point number. The 'b' part indicates that the last b card columns are to be interpreted as following the decimal point unless a decimal point is actually punched. For example, (F10.6) means: a floating point number, occupying 10 card columns and having 6 decimal places. However, if the number 66.666ΔΔΔΔ is punched in these 10 columns it will be read as 66.666, whereas if the number 1023456789 is punched the decimal point will be implied as falling between the '3' and the '4' to give 6 places of decimals, i.e. 1023.456789;

(vii) Format labels must be whole numbers in the range 1 to 99999 (less than this on some computers). The label is punched in columns 1–5 and the rest of the statement in columns 7 to 72, inclusive (though all of these columns may not be needed in any particular case).

Warnings

(i) Never punch commas to separate groups of digits in long numbers. E.g. the value 'one million' would be punched as 1000000 and *not* as 1,000,000.

(ii) Do not specify formats which indicate that more than 80 columns are to be punched —a card has only 80 columns. E.g. (10F12.4) is not a correct format specification, because only 6 twelve-digit numbers can be punched on one card. The correct specification would be (5F12.4), reading 5 numbers from each of two cards.

(iii) Never specify an F code if the number to be read is to be stored in an integer location; conversely, do not specify an I code if the number is to be stored in a floating point location.

(iv) Remember to separate format descriptors by commas. E.g. (10F4.1,I6) is correct, whereas (10F4.1I6) is not.

(v) Do not confuse the digit 0 and the letter O when punching data or programs.

SUMMARY OF OUTPUT FORMAT INFORMATION

(i) Each number to be printed is associated with a set of character positions.

(ii) Floating point numbers require the format descriptor F. Integers require the format descriptor I.

(iii) The use of format specifications is essentially the same as for input formats except that when F descriptors are used the decimal point is counted as occupying one character position. Thus, the number 10.75 would be printed as

Character Position:	1	2	3	4	5
Digit:	1	0	.	7	5

if the F5.2 format descriptor was used. Similarly, the minus sign occupies one character position which must be allowed for in the format descriptor.

(iv) Output format specifications must begin with carriage control characters. These are:

1HΔ, start printing on next line of printout
1H0, leave a line before printing
1H1, start printing at the top of a new page

(v) Each line of output can hold 120 characters. (Some printers have a line-length of 132 or 160 lines.) An output format must not specify more than 120 (or 132 or 160) characters per line.

(vi) Format labels must be integers in the range 1–99999[1]. Labels are punched in columns 1–5[2] and the rest of the statement in

[1] Less than this on some computers.
[2] 2 to 5 on some computers.

columns 7–72 inclusive. Columns 6 and 73–80 are not used (though their use will be described later).

3.3 Basic Arithmetic Operations

While it is occasionally useful to have a program that simply reads in and prints out a set of numbers, as for example in the production of tables, most computer programs are required to carry out some calculations on the numbers read in as data and then to print out the results. A simple example (used later in this section) involves the reading in of the populations of several counties and the computation of the mean population, and other examples were given in Section 3.2. Fortran arithmetic is superficially very similar to 'normal' arithmetic. Statements of the type $6+5=11$ (interpreted verbally as 'the sum of 6 and 5 is eleven') are familiar to most of us, and some of us might recall, maybe with horror, those algebraic equations such as $z=x+y$ that we met at school. Here, z, x and y represent quantities and the statement $z=x+y$ defines the restrictions on the values x, y and z can take. Fortran arithmetic statements are **not** the same as algebraic or arithmetic statements of the form just mentioned, although they are similar in appearance. The signs for the operations of addition and subtraction in Fortran are + and − respectively, but the statement

```
C = A + B
```

does not mean 'C equals A plus B'.

C, A and B are the names or identifiers of words within the central store. The statement 'C equals A plus B' is ridiculous if C, A and B are names of words and not values of quantities. The statement actually means: replace the value currently held in word C by the value obtained by adding the numbers held in words A and B. The lesson is: do not confuse the name of a word with the value stored in that word. The operator '=' means 'replace the contents of the word to the left of the = sign with the value of the expression to the right of the = sign'. It *does not* mean 'equals'. To illustrate this point, consider the following situation:

C	A	B
101.724	60.0	30.0

The values held in C, A and B are 101.724, 60.0 and 30.0 respectively. The Fortran statement C = A + B involves

(1) the setting of an accumulator (Section 2.4(3)) to zero;

(2) addition of the contents of A and the contents of the accumulator and replacement of the contents of the accumulator by this sum;

(3) addition of the new contents of the accumulator and the value contained in word B;

(4) replacement of the value of C by the value obtained at stage (4).

This is shown in Figure 3.1, which shows that (a) the contents of words A and B are un-

Figure 3.1

	Accumulator	A	B	C
Step 1	0.0	60.0	30.0	101.724
Step 2	60.0	60.0	30.0	101.724
Step 3	90.0	60.0	30.0	101.724
Step 4	90.0	60.0	30.0	90.0

changed by the statement, and (b) the original contents of C are lost ('overwritten').

The following short program (Table 3.3)

Table 3.3

```
      READ(5,1) HT1,HT2,HT3
    1 FORMAT(3F10.0)
      C=HT1+HT2+HT3
      WRITE(6,2) C
    2 FORMAT(1H0,F10.4)
      STOP
      END
```

evaluates the sum of three floating-point numbers stored in locations HT1, HT2 and HT3; these might represent the heights of three boys. The statement on the first line causes three floating point numbers to be read from the first data card and placed in the locations called HT1, HT2 and HT3. (You need not worry about the naming of locations—the operating system will set up three locations for your use and will keep a 'map' in the computer's memory to show itself where these locations are.) The second line, which contains the statement labelled 1, tells the operating system that these three numbers, which are floating point in type, occupy columns 1–10, 11–20 and 21–30 of the data card. If no decimal point is punched it will be assumed to follow the last column of each input field, i.e. after columns 10, 20 and 30 respectively. (Remember that blanks within these 10-column fields are interpreted as zeros.) The statement on the third line carries out the summation in the manner already described, and the result is printed by the statement on the fourth line in accordance with the format specified on the fifth line. This output format causes one floating point number to be printed following a blank line (1H0). The number will have ten character positions allocated to it; the last four will be the decimal fraction, with the decimal point printed in position 5. Zeros before the first significant figure before the decimal point are omitted. The instruction 'STOP' tells the computer to stop calculating, while 'END' indicates that the seventh line is the final line of the program.

Multiplication and division are indicated by the signs * (multiply) and / (divide). Thus, PROD = X * Y means: replace the value stored in location PROD by the product of the values to be found in locations X and Y. Multiplication and division are carried out before addition and subtraction so the statement Z=X+V / 10.0 would cause the value in location V to be divided by 10.0 before being added to the value in X, the eventual result replacing the contents of Z. If the value stored in X is 5.0 and the value in V is 32.0 then after the execution of this statement the value in Z would be $5.0+32.0/10.0=5.0+3.2=7.2$.

It would be reasonable to ask: how can I get the computer to evaluate $5.0+32.0$ all divided by 10.0? The answer lies in the use of brackets or parentheses. To cause the addition to precede the multiplication we would write

```
PROD = (X + V)/10.0
```

and the contents of PROD after execution of this statement would be 3.7 (and not 7.2 as before). The rule is: bracketed expressions are evaluated first, although within the brackets the order of precedence is still multiply and divide before add and subtract. Where two operations have equal precedence the left hand one is executed first. Any number of pairs of brackets may be used; too many do no harm as long as they are matched, but too few can result in erroneous answers.

SUMMARY OF FORTRAN ARITHMETIC

1. The '=' sign means 'replace' not 'equals'.
2. Symbols for the arithmetic expression are + (add), − (subtract), * (multiply) and / (divide).
3. * and / operations precede + and −.

4. Where operations are of equal precedence they are executed in order from the left.
5. Brackets alter this order of precedence; expressions within brackets are evaluated first, though the rules contained in notes 3 and 4 above still apply within the brackets.
6. Any number of pairs of brackets is permitted. (A common error is to have too many opening or too many closing brackets.)

Example Program 3

Find the mean of the numbers 3.2, 4.6, 7.5 and 3.0.

Table 3.4

```
   AMEAN = (3.2 + 4.6 + 7.5 + 3.0)/4.0
   WRITE (6,1) AMEAN
 1 FORMAT (1H0,F10.3)
   STOP
   END
```

The answer (4.6) is printed out as

ΔΔΔΔΔ4.600.

Notice that the expression to the right of the '=' on line 1 of the program uses brackets to ensure that summation precedes division. The result of AMEAN $=3.2+4.6+7.5+3.0/4.0$ is 16.05 not 4.6, because this expression is evaluated as $3.2+4.6+7.5+(3.0/4.0)=3.2+4.6+7.5+0.75=16.05$. This program is rather limited in that it will only find the mean of four given numbers; let us now write a program to compute the mean of any four numbers.

Table 3.5

```
   READ(5,1) A,B,C,D
 1 FORMAT(4F5.0)
   AMEAN=(A+B+C+D)/4.0
   WRITE(6,2) AMEAN
 2 FORMAT(1H0,F10.4)
   STOP
   END
```

A possible data card would be:

ΔΔ3.1 ΔΔ4.6 ΔΔΔΔ1 6.5ΔΔ

The statement on line 1 of this program would cause the values 3.1, 4.6, 1.0 and 6.5 to be stored in locations A, B, C and D respectively. This sum, 15.2, divided by 4.0 is stored in location AMEAN and this value (3.8000) is printed out by the statement on line 4.

EXERCISES

(6) Write down Fortran statements to carry out the following calculations:

(i) $f=6.0+5.0$

(ii) $r=7.2+3.6-\dfrac{2.7}{13.9}$

(iii) $a=\pi r^2$ where $\pi=3.1415926$ and r is the radius of a circle in inches. (a is the area of the circle in square inches, and π is the Greek letter pi.)

(iv) $g=(a-b)^2+(c-d)^2$ (If (a,b) and (c,d) are the grid references of two points then g is the square of the straight-line distance between them.)

(v) $d=s/a$ (If s is the total length of streams in a drainage basin and a the drainage area then d is the drainage density.)

(7) Write a Fortran program to read the values Q and CS, where Q is the discharge of a stream in m^3/sec. and CS the suspended sediment concentration in mg/l, and compute the sediment discharge QS (in kg/sec.) from the relationship

$$QS=\frac{Q\times CS}{1000}$$

(8) Hammond and McCullagh (1974, pp. 144–147) provide the following data for a chi-square test of the hypothesis that the number of farms relying on milk does not vary with geology.

Area	Rock Type	Observed No. of Farms relying mainly on milk (O)	Expected No. of Farms relying mainly on milk (E)
1	Keuper Marl	8.0	6.0
2	Upper Sandstone	10.0	6.0
3	Upper Marl	6.0	6.0
4	Clay-with-Flints	3.0	6.0
5	Culm Measures	3.0	6.0

Write a program to calculate

$$\chi^2 = \Sigma((O-E)^2/E)$$

(Hints: (i) the chisquare test is described in a number of elementary statistics textbooks, e.g. Hammond and McCullagh (1974), Ebdon (1976), Cole and King (1968), Gregory (1963).

(ii) set out your program as follows:

1. Read the floating point values 8.0, 10.0, 6.0, 3.0 and 3.0 into locations O1, O2, O3, O4 and O5.
2. For the values in O1, O2, . . . O5 calculate a chisquare value, for example, CHI1 = (O1 – 6.0)∗(O1 – 6.0)/6.0;CHI2 = (O2 – 6.0) ∗(O2 – 6.0)/6.0 etc.
3. Sum the values in the locations CHI1, CHI2, . . . CHI5. Put the result into location CHISQ.
4. Output the contents of location CHISQ.)

3.4 Identifiers, Arrays and DO Loops

In Sections 3.2 and 3.3 the names (identifiers) of storage locations or words were used with very little comment. There are, however, several easily-learned rules associated with identifiers that ought to be memorised at this stage. They are:

(i) An identifier must begin with a letter.

(ii) It can be up to 6 characters long.

(iii) Characters 2 to 6 (if any) can be any letter or digit.

(iv) The initial letter of the identifier tells the operating system the type of number to be stored in the named location. If the initial letter is I, J, K, L, M or N the number stored is of type integer, but if the initial letter is in the range A–H or O–Z then the number will be stored in floating point form.

The following identifiers could be used to name locations in which integer numbers will be stored: I, J, JOHN, KATE, NUMBER, N01, N005. Floating point numbers would be stored in locations identified as for example X, Y, Z, STDEV, FRED or CHISQ. (Refer to Section 2.4.1. for definitions of the terms 'integer' and 'floating point'.) It is not illegal in some Fortran systems to use the names of Fortran instruction words such as STOP, END, FORMAT as identifiers but it is good practice not to do so.

EXERCISE

(9) State (a) whether the following identifiers are legal and, if so, (b) the type of number that could be stored in them? The first three answers are given.

	Identifier	Legal?	Type
(i)	JULIUS	yes	integer
(ii)	XNEW	yes	floating point
(iii)	0IAM72TODAY	no (– begins with numeral and has too many characters)	
(iv)	FRATIO		
(v)	TVALUE		
(vi)	MEAN		
(vii)	Z005Z		
(viii)	STANDARD		
(ix)	X∗Y		
(x)	BOOK01		

As a further exercise, look back over Sections 3.1 to 3.3 and check the legality and type of the identifiers used there.

In the rest of this section the idea of naming one location will be extended to that of giving the same name to several locations. This idea may appear impracticable or even unnecessary, but a simple analogy will help to explain its importance. Many British students have spent part of their Christmas vacation working for the Post Office as temporary postmen (or women). In the postal service each letter is addressed to a particular location (a house) which might be thought of as being analogous to a location (a word) within the computer's memory. In a place as large as London or Manchester, or even Little Snoring in Norfolk, the postman's job would be an unenviable one if the streets did not have names. Every house would then need to have a unique name, perhaps allocated by the Post Office. This in itself would be both a complicated and unnecessary task. The system of numbering the houses and naming the streets is both simpler and less bureaucratic—a romantic householder can still give his house a name, while a stranger might find it comforting to know that No. 5, Station Villas was close to No. 7, probably next door.

A similar convention can be used within a computer program to avoid naming all the locations used by that program. If we turn back to Exercise 8 we find that 11 separate names were used to identify locations containing the observed values and the chisquare values. It would be far easier to give one name to the set of locations containing the five observed values in the farm example and another name to the set of locations containing the chisquare values. Each location within a set is identifiable by a number recording its relative position. This is shown diagrammatically in Figure 3.2. The location containing the fourth observed frequency (i.e. the number of farms relying mainly on milk on the Clay-with-Flints) is referred to by the identifier OBS(4), and the corresponding chisquare value is stored in CHI(4). These groups of words with a single identifier are called *arrays*. The number within the brackets is the *array subscript*. Array subscripts in Fortran must always be positive integers. Even if a floating point location such as F contains a whole number, e.g. 4.0, we cannot refer to location OBS(F) because F is not an integer name. K is an integer name, so the reference OBS(K) is legal provided that K contains, in this case, one of the values 1, 2, 3, 4 or 5.

EXERCISES

(10) Are the following array references permissible? If not, give reasons. (Assume that each array contains 5 locations.)

(i) A4(4) (ii) 6A(4)
(iii) JOHN(0) (iv) JOHN(10)
(v) ABCD(−1) (vi) JAMES(4.0)

(11) What type of number (floating point or integer) would be contained in the elements of an array named PETER?
(Hint: the rules concerning identifiers apply also to array names.)

How does the computer distinguish between array names and names of single locations? The programmer must tell the operating system which names he is using for arrays by means of a special statement called a DIMENSION statement. This is of the form

Figure 3.2

OBS	1	2	3	4	5	Position
Array name	8.0	10.0	6.0	3.0	3.0	Value

CHI	1	2	3	4	5	Position
Array name	0.66	2.66	0.0	1.50	1.50	Value

DIMENSION *array name* 1 (size), *array name 2* (size), ..., *array name n* (size) and it must always be placed before any executable statement, that is, an instruction which causes the computer to carry out some operation in contrast with a purely informative or non-executable statement. A FORMAT statement is nonexecutable but a READ statement is executable. For example, to reserve two floating point arrays, each of ten words in length, the first statement of the program would be

```
DIMENSIONΔFRED(10),HARRY(10)
```

The statement must be punched between columns 7 and 72 of a punch card. Notice that there is no comma after the name and size of the final array in the line. It is not necessary to use all the space reserved for any array, so the size of the arrays is normally set to the maximum number of locations that will usually be required. This avoids changing the DIMENSION statement every time the program is run.

The provision of arrays in Fortran is in itself not of great significance, but when arrays are used in association with DO loop instructions a great deal of programming effort can be avoided. The DO loop instruction allows for the automatic modification of the array subscripts so that the programmer need not supply explicit numerical values for these. Suppose, for example, an array set up in a DIMENSION statement and called FRED holds a set of 4 floating point numbers and that we wish to add each of the numbers stored in FRED in turn to the values stored in SUM, which is initially zero. The hard way of doing this would be:

```
. . . . . . . . . . . .
SUM = 0.0
SUM = SUM + FRED (1)
SUM = SUM + FRED (2)
SUM = SUM + FRED (3)
SUM = SUM + FRED (4)
. . . . . . . . . . . .
```

The easy way is:

```
  . . . . . . . . . . . .
  SUM = 0.0
  DO 1 I = 1,4
1 SUM = SUM + FRED (I)
  . . . . . . . . . . . .
```

The statement DO 1 I = 1,4 means: 'carry out the instructions following this statement up to and including the statement labelled 1 for values in location I beginning at 1, increasing by one every time the loop is executed and going up to 4'. Since I is an integer name the values stored in I will be: 1, 2, 3, 4. Thus, the DO loop in this example would cause the contents of FRED (I) to be added to the contents of SUM with I taking values 1, 2, 3, 4. When the loop has been executed for I = 4 (i.e. the fourth and last time) control passes to the statement following the one labelled at the end of the DO statement (label 1 in the example).

Certain rules must be followed:

(i) the DO statement must refer to a label number (1 in the above example) which occurs after the DO statement in the program;

(ii) the loop index or counter (I in the above example) must be an integer variable name. It cannot be an array name or an arithmetic expression or the name of a floating point location;

(iii) the lower bound (1 in the example above) of the loop must be a positive integer or integer variable name, and the upper bound (4 in the example above) must be larger than, or at least equal to, the lower bound. The upper bound must also be an integer or an integer variable name.

Example Program 4

Read 10 values into the floating point array X, then multiply each of the 10 values by 2.0 and print out the results. The first line of this program reserves 10 storage locations for the array X which, because the initial (and sole) letter of the array name falls outside the range I–N, holds floating point numbers. Lines 2 and 3 cause 10 values to be read from separate cards in the data part of the input. Initially I is set to 1 and the value read from the first data card is put into location X(1). Then I is incremented (increased) by 1 and the READ instruction is executed a second time, the number read being placed in location X(2). This process is continued until I, serving as a counter, has taken all the integer values in the range 1–10. Notice that every time a READ statement is executed the reading of a new data card is begun. Line 4 specifies the input format, labelled 2. Lines 3 and 4 are equivalent to the 10 statements:

```
X(1) = X(1)*2.0
X(2) = X(2)*2.0
 .       .
 .       .
 .       .
X(10) = X(10)*2.0
```

Lines 7–9 cause the 10 values in the array X to be output to the lineprinter. The WRITE statement is executed 10 times, so 10 new lines are begun (i.e. each number is printed on a separate line).

Table 3.6

```
   DIMENSION X(10)
   DO 1 I=1,10
 1 READ(5,2) X(I)
 2 FORMAT(F10.0)
   DO 3 I=1,10
 3 X(I)=X(I)*2.0
   DO 4 J=1,10
 4 WRITE(6,5) X(J)
 5 FORMAT(1H ,F10.4)
   STOP
   END
```

Example Program 5

This program is a modification of Example Program 4. It allows the number of data values to be read into array X to be specified along with the data themselves. This is a most useful feature for it enables the programmer to specify only the maximum rather than the actual size of the array (X in the example) in the DIMENSION statement. The first data card will contain an integer number punched in columns 1–4 (14 format). This number will be stored in the integer location N. Thereafter, any reference to N will cause the value stored in N to be substituted. Since the size of the array X declared in the DIMENSION statement is 250 the value of N must not exceed 250 or else the program will fail, and an error message or error number will be printed out (Refer to Appendix C for details). The rest of the program is similar to Example Program 4.

Table 3.7

```
   DIMENSION X(250)
   READ(5,6) N
 6 FORMAT(I4)
   DO 1 I=1,N
 1 READ(5,2) X(I)
 2 FORMAT(F10.0)
   DO 3 I=1,N
 3 X(I)=X(I)*2.0
   DO 4 J=1,N
 4 WRITE(6,5) X(J)
 5 FORMAT(1H ,F10.4)
   STOP
   END
```

Example Program 6

At a time of increasing raw material costs, punching one number per card and outputting one number per line on the paper print out seems a little wasteful as a punch card has 80 columns and a line of print can accommodate 120 or more characters. We will therefore rewrite Example Program 5 to make it more economical in terms of data cards and lineprinter paper.

The statement READ (5,2)Δ(X(I),I = I,N) is interpreted in exactly the same way as the combination of the two statements DO 1 I = 1,N and 1 READ (5,2)ΔX(I) except that the READ statement is executed once only and the FORMAT statement labelled 2 is the only control over the number of values entered on each data card. If the value 40 is punched on data card 1, directing that the value 40 be placed in location N, then the four subsequent data cards will contain 8 values each (format (8F10.0)) with each value occupying 10 card columns. Similarly, the WRITE statement on line 7 is executed once only, and the number of characters printed on every line is controlled by the FORMAT statement labelled 5.

The two FORMAT statements labelled 5 and 2 have been placed after the STOP statement to show that they can be located anywhere in the program between the DIMENSION and END statements. They do not necessarily have to follow the READ or WRITE statements which reference them.

Table 3.8

```
      DIMENSION X(250)
      READ(5,6) N
    6 FORMAT(I3)
      READ(5,2) (X(I),I=1,N)
      DO 3 I=1,N
    3 X(I)=X(I)*2.0
      WRITE(6,5) (X(I),I=1,N)
      STOP
    2 FORMAT(8F10.0)
    5 FORMAT(1H ,12F10.4)
      END
```

EXERCISES

(12) Write a program to read the number of data values into an integer location N, then read the N data values into the elements of an array OBS. Find the sum and the average value of the elements of OBS. (Hint: start with

```
      DIMENSION OBS (250)
      READ (5,1) N
    1 FORMAT (I3)
      READ (5,2)(OBS(I),I = 1,N)
    2 FORMAT (10F8.0)
      .................. )
```

(13) Use the program you wrote in Exercise 12 to find the mean population of the following towns:

Barrow	67,000	Blackburn	111,000
Blackpool	147,000	Bolton	167,000
Bootle	75,000	Burnley	85,000
Bury	59,000	Liverpool	789,000
Manchester	703,000	Oldham	121,000
Preston	119,000	Rochdale	88,000
Salford	178,000	St. Helens	110,000
Southport	84,000	Warrington	81,000
Wigan	85,000	Lancaster	52,000

3.5 Conditional Statements, GO TO statements and ABS Function; Mean and Variance; Newton's Square Root Algorithm

The purpose of this section is to consolidate the knowledge gained in the first four sections of this chapter, and to add some more Fortran instructions to your vocabulary. This will be done by writing and testing two example programs. The first program will find the arithmetic mean and variance of a sample of measurements. The second program introduces the idea of the algorithm or computing formula.

Example Program 7

The mean of a set of n observations on a variable x_i ($i=1, 2, 3, \ldots, n$) is calculated by summing the x values and dividing by n. The formula is

$$\bar{x}=\sum_{i=1}^{n} x_i$$

('x bar equals sigma x – i'). This is equivalent to the formula

$$\bar{x}=(x_1+x_2+x_3+\ldots.+x_n)/n$$

where x_1 is the value of the first observation, x_2 the value of the second, and so on. The sigma sign $\sum_{i=1}^{n} x_i$ means simply: add together the x values following the sigma sign, i.e. find $x_1 + x_2 + \dots + x_n$.

The variance of the measurements is an indication of the scatter of the observed values around the mean value. It is calculated by summing the squared differences between each observed value and the mean value, and dividing this sum by n, that is

$$\text{variance of } x = \sum_{i=1}^{n} (x_i - \bar{x})^2 \div n$$

or

$$s_x^2 = \sum_{i=1}^{n} (x_i - \bar{x})^2 / n$$

The value s_x^2 is the variance of a sample of size n drawn from a given population[1].

The computer program is set out in Table 3.9. As an exercise, go through it line by line and write down a commentary on each instruction.

EXERCISES

(14) Why are the contents of XBAR and VARX set to 0.0 and not simply 0 on lines 4 and 5?

(15) How many values are punched on each data card? In what columns?

(16) Explain the action of the DO loops terminating at lines labelled 2 and 3.

(17) On line 8 a floating point value, the contents of location XBAR, is divided by the number of observations, 50. Since XBAR is a floating point location we must divide its contents by the floating point equivalent of the integer value 50, that is, 50.0. Similarly on line 11 the contents of VARX are divided by 50.0, not 50, which is an integer. If any given integer value is stored in an integer location such as N, its floating point equivalent is found by replacing the contents of any available floating point location such as F by the value FLOAT (N). For example,

```
   FVAL = FLOAT(K)
or RUM = FLOAT(MUM)
or VARX = VARX/FLOAT(N)
```

Table 3.9

```
      DIMENSION X(50)
      READ(5,1) (X(I),I=1,50)
    1 FORMAT(10F8.0)
      XBAR=0.0
      VARX=0.0
      DO 2 I=1,50
    2 XBAR = XBAR + X(I)
      XBAR = XBAR/50.0
      DO 3 I=1,50
    3 VARX=VARX + (X(I) - XBAR) * (X(I) - XBAR)
      VARX = VARX/50.0
      WRITE(6,4) XBAR, VARX
    4 FORMAT(1H ,2F10.4)
      STOP
      END
```

[1] See Ebdon (1976) for a discussion of sample variance, population variance and best estimate of population variance.

Using this instruction, modify Example Program 7 so that any number n of observations on the variable x can be read in, up to a limit of 600.

(18) Modify the program written as Exercise 17 so that (a) 10 floating point values, each occupying 8 card columns, are read from each data card, and (b) the values in XBAR and VARX are printed on separate lines. Try to answer part (b) of this problem without changing the WRITE statement.

(19) Use Example Program 7 to calculate the (unweighted) mean and variance of the percentage of the populations of the following towns in the age group 65 or more:

Brighton	15.4	Eastbourne	19.0
Hastings	19.7	Hove	21.3
Worthing	24.6	Torquay	17.9
Bournemouth	18.9	Southport	17.5
Blackpool	15.2	Harrogate	16.0

Now find the (unweighted) mean and variance of the following set of measurements, which are also percentages of total population aged 65 or over, and compare your results with those obtained for the holiday resorts:

Cardiff	9.8	Sheffield	10.4
Oldbury	7.6	Coventry	8.0
West Bromwich	8.3	Newport	9.4
Huddersfield	11.9	Dudley	8.8
Stoke-on-Trent	8.5	Nottingham	9.9

Example Program 8

The square root of a number is sometimes required in statistical programs. In Program 7 of this chapter we could have decided to calculate the sample standard deviation s_x of x as well as the variance. (The standard deviation is the square root of the variance.) Actually, a very simple Fortran instruction exists for finding square roots, but before telling you what it is in Section 3.6 a method of finding the square root of a given number using a method developed by Isaac Newton will be described. Newton gave an exhaustive set of instructions which would, if followed exactly, produce a close approximation to the square root of a given positive real number. Such a set of instructions is called a computing algorithm or simply an algorithm. The reduction of a problem such as 'find $\sqrt{2}$' to the form of an algorithm is necessary before a program can be written. Newton's algorithm is as follows.

(i) Call the given number a and its square root x, i.e. $x^2 = a$ or $x = \sqrt{a}$.

(ii) Guess the value of x, calling this first approximation x_1 ('x-one' or 'x subscript one').

(iii) A better estimate, x_2, of $\sqrt{a}$ is now given by the following formula:

$$x_2 = \tfrac{1}{2}(x_1 + \frac{a}{x_1})$$

(iv) If the difference between x_2 and x_1 is small, that is if convergence has occurred, the algorithm terminates, and x_2 is very close to $\sqrt{a}$. Otherwise repeat step (iii) to determine a third approximation x_3, and carry out the convergence test of step (iv). Repeat steps (iii) and (iv) until the convergence criterion is satisfied. As an example let $a = 100$ and take $x_1 = 100$. (Steps (i) and (ii)). 100 is obviously a poor guess of the value $\sqrt{100}$, but we will pretend ignorance. To find a better approximation we take the formula of step (iii) and calculate

$$x_2 = \tfrac{1}{2}(100 + \frac{100}{100}) = 50.5$$

The difference (step (iv)) between x_1 and x_2 is 49.5, which is not small in comparison with the value of a, so we repeat step (iii) and find

$$x_3 = \tfrac{1}{2}(50.5 + \frac{100}{50.5}) = 26.24$$

The difference (50.5 − 26.24) is still large relative to $a = 100$, so a fourth estimate x_4 is computed:

$$x_4 = \tfrac{1}{2}(26.24 + \frac{100}{26.24}) = 15.025$$

We need not continue with these calculations to show that x_i is converging on 10.0, the square root of 100. A repetitive series of calculations such as Newton's algorithm is called an *iterative procedure*. Not all iterative procedures are guaranteed to converge (i.e. produce a result that is close to the correct value) and convergence may be slow even with a good algorithm if the convergence criterion (the maximum acceptable difference between the results of two successive iterations) is set too small. It would not, for example, be sensible to continue Newton's algorithm until the relative difference between successive estimates was 10^{-50} (10^{-50} is 49 zeros and a one following the decimal point). A good value for the purposes of this exercise would be 10^{-4} ($=0.0001$).

One point that we have not covered is: what happens if our initial guess, x_1, is less than $\sqrt{a}$? In that case the value $x_2 - x_1$ would be negative, and a negative value would by definition be less than $+0.0001$. The answer is that the *absolute* value of the difference between successive estimates, rather than the actual value, should be used. The absolute value of a quantity is the value of that quantity ignoring the sign. The absolute value of a quantity b is written: $|b|$. If $b=6$, $|b|$ also equals 6. If $b=-0.5$, $|b|=0.5$. If $b=0$, $|b|$ also equals 0. In Fortran, the absolute value of a floating point quantity held in a location B is given by the value of ABS(B). For example, the instruction

```
B = ABS(X)
```

will replace the current contents of word B with the value held in X with the negative sign (if any) ignored. If the absolute value of an integer quantity is required the instruction is

```
K = IABS(M)
```

where both K and M are the names of integer locations. It must be remembered that words such as ABS, IABS, FLOAT, DO, DIMENSION, FORMAT, STOP and END are Fortran instruction words and have a special meaning. In contrast, the names of locations such as SUM, X, OBS or I are used by the programmer to identify those store locations that he is using in his program. Provided that they comply with the rules given in Section 3.4 these names can be chosen at will.

Newton's algorithm can be easily programmed using the ABS function. The values of a and x_1 must be read in or supplied, and successive values of x computed using the relationship of step (iii) which can be generalised to:

$$x_{i+1} = \tfrac{1}{2}(x_i + a/x_i) \qquad (3.1)$$

and the convergence criterion of step (iv) which can also be generalised to:

$$\frac{x_{i+1} - x_i}{x_{i+1}} < \varepsilon \qquad (3.2)$$

The $<$ sign means 'is less than', and ε (Greek epsilon) is the value of the convergence criterion, which will be taken to be 0.0001.

One more Fortran statement is needed before the program can be begun. Step (iv) of the algorithm states that if the relative difference between successive approximations is less than ε, then either the algorithm terminates or step (iii) is repeated. This can be written in a shorthand form as:

if (condition satisfied) then execute instruction.

For example, the condition in this case is given by equation (3.2); if it is satisfied, the instruction to be executed is: end the iterative procedure, print the results, and terminate the program. If it is not satisfied, then perform one more iteration. In Fortran the characters .LT. stand for 'is less than' in a conditional statement of this sort. For example, the condition might be stated as: if x is less than 0.0001 then end the iterations. It could then be programmed as:

```
IF (X.LT.0.0001) GO TO 4
```

where 4 is the label of the statement following the section of program in which the iterations are performed. GO TO statements can come anywhere in a Fortran program, so long as they lie between the DIMENSION statement (if any) and the STOP statement.

Conditions involving logical comparisons other than 'less than' are permitted. The full list of operators is:

.GT.	is greater than
.EQ.	is equal to
.NE.	is not equal to
.LE.	is less than or equal to
.GE.	is greater than or equal to
.LT.	is less than

The . (decimal point) on either side of the operator must not be omitted. Examples of the use of these logical operators are:

```
IF (X.NE.Y) X = 0.75*X
IF (X.EQ.Y) STOP
IF (I.EQ.2) GO TO 10
IF (A*B + 2.7 .LT. ALIM) WRITE (6,7)A,B
IF (I.EQ.1) READ (5,107)A
```

Further examples of conditional statements and logical operators are given in Section 4.2.

We will now rewrite the algorithm using a notation more suited to programming so that the form of the conditional statement is seen more clearly.

S1: read x_1 and a, then set i to 1
S2: find a new value x_{i+1} using equation (3.1)
S3: if the condition given by equation (3.2) is satisfied go to step 4 otherwise go back to step 2 having increased the value of i by 1
S4: print x_i, a and i
S5: stop

Table 3.10

```
      READ(5,1) X,A
    1 FORMAT(2F10.0)
      I=1
    2 XNEW=0.5*(X+(A/X))
      CONV=ABS((XNEW-X)/XNEW)
      IF(CONV.LT.0.00001) GOTO 4
      I=I+1
      X=XNEW
      GOTO 2
    4 WRITE(6,100) XNEW, A, I
  100 FORMAT(1H0,2F11.5,I4)
      STOP
      END
```

The conditional instruction is on line 6. It states that, if the value held in the floating point location CONV is less than 0.0001 then the next instruction to be executed is that labelled 4, which is the first statement of step S4; otherwise the next instruction to be executed is that on the immediately following line (line 7). The statement on line 5 illustrates the use of the ABS function; it causes the difference, with the sign ignored, between XNEW and X divided by XNEW to be placed in location CONV.

The program logic is also interesting. Instead of providing a fresh location for each successive approximation only two locations are used. Location X contains the value of the approximation at the previous iteration and XNEW is the value computed from (3.1) for the current iteration. If the condition of line 6 is not satisfied, the contents of X are replaced by those of XNEW (line 8) and the program returns to label 2 (line 4) to calculate the value of the new contents of XNEW. The value held in integer location I, initially set to 1, is increased by 1 at every iteration; when the process converges, location I contains the value of the number of iterations completed.

To illustrate the working of the program we will simulate its action over three iterations, showing how the contents of X, XNEW, CONV and I alter (Table 3.11). The values of A and X will be the same as in the example, that is, each is 100.0. The symbol ? indicates that the contents of the location are not yet defined.

EXERCISES

(20) Complete Table 3.11 for iterations 3 and 4.
(21) Write out a worked example, following Table 3.11, of Newton's algorithm which computes $\sqrt{25.0}$ using a starting value of 2.0. Give a careful explanation of the need for the use of the ABS function on line 5 of the program.
(22) Are all the brackets on the fourth and fifth lines of the program necessary? Can you rewrite these lines using fewer brackets?
(23) Insert an instruction to print the values held in locations XNEW and I at the end of each individual iteration cycle.

3.6 Captions, More Functions; Product-Moment Correlation Coefficient

So far we have considered only input and output formats for numbers, either of type integer (I) or floating point (F). If a program involves the output of many numerical results, labelling them with (alphabetic) names would help in interpreting the printout. For example, CHISQUARE = 10.62 WITH 10 D.F. is easier to comprehend than 10.62 10. Although there are several alternative ways of dealing with alphabetic information, we will consider only two in this section. The first of these is the 'A' format, which allows alphanumeric data (alphabetic + numeric + special symbols) to be read from cards and transferred to the computer's memory, or vice-versa—an 'A' format can be used to transfer alphanumeric infor-

Table 3.11 First Three Iterations in Square Root Algorithm

I	X	XNEW	CONV	A
1	100.00	50.50	0.9801	100.00
2	50.50	26.24	0.9245	100.00
3	26.24	15.03	0.7461	100.00
.	.....	.....	......	......
.	.....	.....	......	......
.	.....	.....	......	......
(6	10.03	10.00	0.0030	100.00)

mation from the central memory to an output device such as the lineprinter. One important use of the 'A' format is the input and output of job identification information, as will be see in the first example.

In Section 2.4(1) the make-up of a computer 'word' or store location was discussed. A word is a group of memory units, each capable of holding one binary digit or bit. The number of bits per word, or word length, varies from one computer to another. A CDC 7600 computer has 60-bit words, while an ICL 1906A has 48 bits per (double) word. Thus, the number of alphanumeric characters that can be held in one word varies from one make of computer to another. All computers can hold a minimum of 4 alphanumeric characters per word, so in the examples that follow no more than 4 characters will be stored in each word. If the computer that you are using is capable of holding more than 4 characters per word, all that will happen will be that more store locations will be used than would be the case if the program had been written specially for that machine. (A small number of computers are exceptional in that only 2 characters are stored in each integer location, though 4 characters are stored in a floating point location.)

In the specification of an A type format all we need specify is the number of characters to be held per store location. This, we have just decided, will be 4 at most. Hence all our A formats will be of the type (nAm) where n specifies the number of locations to be used, and m the number of characters held in each location, which will be 4, 3, 2 or 1. For example, an input format may be required to read in 80 alphanumeric characters from a data card. If we decide to store 4 characters per location we find that 80 divided by 4 is 20, so 20 locations will be required to hold the contents of the data card. The A format specification will therefore be (20A4).

Example Program 9

Write a program to read in 80 alphanumeric characters into an array ALPHA, storing 4 characters per word[1] (Table 3.12). One could

Table 3.12

```
      DIMENSION ALPHA(20)
      READ(5,100) (ALPHA(I),I=1,20)
  100 FORMAT(20A4)
      STOP
      END
```

store less than 4 characters per location; if it was required to put only one character into each location then the example program would be as shown in Table 3.13.

Table 3.13

```
      DIMENSION ALPHA(80)
      READ(5,100) (ALPHA(I),I=1,80)
  100 FORMAT(80A1)
      STOP
      END
```

Output of alphanumeric information causes no problems, provided we remember that (a) all output formats begin with a carriage control character (1HΔ, 1H0, or 1H1) and (b) the output format specification must not require that more or less characters are read from a location that were input to it. For example, if two characters were stored in each of the 20 locations forming the array ALPHA in Example Program 9 and on output it was specified that four characters were contained in each location, then the excess characters would be printed as blanks.

The storage mode used to hold alphanumeric information can be either integer or floating point, provided that the same type is used throughout the program. For instance, if the integer location IVY contains the characters ABCD then after the execution of the following instruction

```
IOAK = IVY
```

[1] Any acceptable identifier would serve as well as ALPHA.

the characters ABCD will overwrite whatever was stored in location IOAK. But, if the instruction had been

```
OAK = IVY
```

then, since OAK is the name of a floating-point location (it begins with the letter O), it would not necessarily contain the characters ABCD. In fact, it would be highly unlikely to do so.

We do not always want to have to read in the alphanumeric information that we want to print out on our output, for some such information is constant while other alphanumeric information might vary from one run of the program to another. Thus, the printout from a program might contain the following alphanumeric information:

```
HENRY SUTCLIFFE DISSERTATION—↔
↔DATA SET 1 (PADDINGTON)
CHISQUARE = 10.62
D.F.      = 10
```

Line 1 of this output will vary from run to run; the next time Henry Sutcliffe runs the program he may want his printout to begin

```
HENRY SUTCLIFFE DISSERTATION—↔
↔DATA SET 2 (KENNINGTON).
```

On the other hand, the phrases 'CHISQUARE =' and 'D.F. =' on lines 2 and 3 will be output in exactly the same manner every time the program is run. While Henry Sutcliffe would have to read in the information to be printed on line 1 at every program run, he would expect to be able to fix the contents of lines 2 and 3 within the program itself. This can be done by means of the H format. H, incidentally, stands for Hollerith, the name of the man who invented the card sorting machine first used in the U.S. census of 1890. H formats are extremely simple to use. Imagine that a programmer wants to output the statement BEWARE OF WELSH GEOMORPHOLOGISTS at the head of his program output. First of all, count the number of characters, including spaces, in this statement (32) and secondly add 1 for the carriage control character (Δ,0,1). The appropriate H format specification would be:

```
(33H0BEWARE OF WELSHΔ↔
↔GEOMORPHOLOGISTS)
```

To print out this statement we need a WRITE instruction, and we need to label the format statement. An example is:

```
      WRITE (6,121)
121   FORMAT (33H0BEWARE OF WELSHΔ↔
      ↔GEOMORPHOLOGISTS)
```

Any label other than 121 could have been used, providing that it is not used as a label anywhere else in that program. Note that the 0 following the 33H is the carriage control character.

A, F and I format specifications can be mixed with H format specifications; for example, you may wish to number the pages of your output using an integer variable KOUNT to store the current page number. Assuming that KOUNT, initially holding a zero, is increased by 1 every time a new page is printed then the following statements could be included within your program:

```
      WRITE (6,97) KOUNT
 97   FORMAT (12H1PAGE NUMBER, I4)
```

A line of lineprinter output cannot hold more than a certain number of characters, usually 120, 132 or 160. Thus an H format specifying 175H0. . . . . would be unacceptable if only 120 characters could be printed on one line. The obvious thing to do is to split the statement to be printed into sections of, at most 120[1] characters and separate each such section in the format specification with a slash (/) character (the same as the 'division' character). This would cause the second part to be printed on a new line. The statement

```
161   FORMAT (175H0 . . . . . )
```

should be written as

```
161   FORMAT (120H0 . . . . . . . / 56H0↔
      ↔. . . . . . . . )
```

[1] Some lineprinters allow the use of lines of 132 or 160 characters long.

Why 56H0, since 120 + 56 = 176 and not 175? The answer is that each line of printer output must start with a carriage control character. Thus, if the second section began with the characters AS LONG AS WE ARE TOGETHER ...... then 55HAS LONG AS ..... would imply that A was the carriage control character. This might result in the phrase being printed as S LONG AS WE ARE TOGETHER, or it could cause the lineprinter to print portions of this phrase randomly. To avoid this inconvenience, always ensure that each H format specification begins with a 0, 1, or Δ, and remember that a new line of printer output follows each / within the format specification.

The next question that requires answering is this: how can a program statement containing more than 66 characters be punched, since a program statement begins in column 7 and cannot proceed beyond column 72? The answer to this is the use of continuation cards. You may have wondered why column 6 of a program statement card did not have a use; in fact, if any character other than the digit zero (0) is punched in column 6 of a program card then the compiler interprets it as meaning that the card contains an instruction that is a continuation of the instruction on the preceding card. It should be stressed that continuation cards can be used for any lengthy Fortran statement, not just those involving H formats. There is an upper limit to the number of continuation cards that can be used at one time, but this is unlikely to be reached in simple programs. Also, it is not necessary to use all of columns 7–72 of the card before going on to a continuation card. It is quite legal to write:

```
      WRITE (6,72)
   72 FORMAT (8H0THIS IS,
     117HΔAΔWASTE OF CARDS)
```

which would be printed as:

```
THIS IS
A WASTE OF CARDS
```

In a long FORMAT specification, however, it is good practice to split up the sections and punch them on separate cards. The resulting program listing is then much easier to read and if, as is possible with a long statement, mistakes have occurred, then subsequent repunching will be minimised.

Incidentally, columns 73–80 can be used to provide card identification. If the program is designed to compute a correlation coefficient the characters COREL001, COREL002, etc. can be punched in columns 73–80 of successive cards. This may be a tedious thing to do but it could save a lot of time if you, or (more rarely) the computer operator, drop the cards. Comments within the program listing can be inserted by punching the letter C in column 1 of a program statement card. All programs including those you write solely for your own use should include sufficient comments to allow another programmer to determine what is happening. It is conceivable that you might wish to modify the program at a later stage. If you cannot remember how it works, this may prove difficult.

To end this section, the use of functions will be placed on a more formal basis. The instructions ABS, IABS and FLOAT have been used without much comment on earlier pages. ABS and IABS cause the absolute value (irrespective of sign) to be computed, while FLOAT will convert an integer value to its floating point equivalent. These are two examples of library functions, which are provided by all standard Fortran compilers. Other functions that are of value in statistical programs are: SQRT, ALOG, ALOG10, EXP, SIN, COS and INT. Their meaning is given in Table 3.14 (in which ABS, IABS and FLOAT have been included for completeness).

Table 3.14 Fortran Library Functions

Function Name	Operation
SQRT	Square root
ALOG	log base e
ALOG10	log base 10
EXP	e to the power of
SIN	sine

Table 3.14 (continued)

COS	cosine
INT	integer part of
ABS	absolute value of (floating point)
IABS	absolute value of (integer)
FLOAT	floating point equivalent of

The name of the function is treated just like the name of any store location. In other words, if the name begins with a letter in the range I to N the function holds (returns) an integer value; otherwise it returns a floating point value. Function names can be used in assignment statements as if they were variable or array names, e.g.:

```
Y = SQRT(X)
Z = ALOG10 (6.3)
L = INT(Z)
```

The name or the value within the brackets (called the argument) must, however, refer to a value which is of a specified type. These specifications are given in the next paragraph. As an example, the argument to SQRT must be of type floating point; consequently the statement Z = SQRT(4) will not necessarily put the value 2.000 .... into location Z. Some operating systems will halt the program at this point and print out an error message, while more flexible operating systems will manage to guess that you really meant Z = SQRT(4.0). It would be very unwise to rely on the operating system to iron out errors in your program.

Details of the library functions given in Table 3.14 are:

SQRT: returns the square root of a floating point argument

Examples: Y = SQRT (Z)
X = SQRT (4.0–Z + 6.7)
A = SQRT (81.9)

ALOG: returns the natural logarithm (base e) of a floating point argument

Example: A = ALOG (3.14159)
Z = ALOG (B*C)

ALOG10: returns the logarithm (base 10) of a floating point argument

EXP: returns the value of e to the power of a floating point argument. This is useful in finding the antilogarithm of a given log base e. For example, if we want to find the number that has the logarithm (base e) of 3.725 then

Z = EXP (3.725)

will cause the antilog of 3.725 to be put into locaton Z.

SIN: gives the sine of an angle expressed in radians. To convert degrees to radians, note that $360° = 2\pi$ radians where π (pi) = 3.15926535897. Therefore x degrees = $(360/2\pi)$ x or $(180/\pi)$ x.

Example: Find the sine of 74°

```
PI = 3.15926535897
XDEG = 74.0
XINRD = XDEG * (180.0/PI)
SINEX = SIN (XINRD)
```

COS: returns the value of the cosine of an angle expressed in radians. (Refer to the previous paragraph for details of converting from degrees to radians.)

INT: returns the integer value of the whole number which is nearest to (but less than) the value of the floating point argument. If the location X holds the value 76.729 then INT(X) returns the value 76 (as an integer).

Example: Find the nearest integer to the floating point value 6.72, held in location FL.
From the discussion above we note that INT (6.72) returns the value 6, whereas the nearest integer to 6.72 is 7. By adding 0.5 to the argument we can find the nearest integer:

NRINT = INT (FL + 0.5)

If the value stored in FL lay between 5.5 and 6.4999 then INT (FL + 0.5) would return the value

6. If the stored value was in the range 6.5 to $7.499\dot{9}$ then INT (FL +0.5) would return the value 7.

ABS: gives the value of the floating point argument irrespective of sign. In other words if the argument is positive or zero it is unchanged by ABS; if it is negative, ABS will remove the minus sign. If the argument is an integer then use IABS in place of ABS.

Examples: X = −7.25
X = ABS(X)
(The value stored in X is now +7.25.)
J = −55
J = IABS(J)
(The value stored in J is now +55.)
X = 10.5
X = ABS(X)
(The value stored in X is unaltered by ABS.)

FLOAT: converts an integer argument to its floating point equivalent. Effectively '.0' is added to the integer, e.g. 7 becomes 7.0.

Examples: X = FLOAT(I)
S = FLOAT(4)

You will have noticed from the examples that all the functions except INT and IABS must be assigned to a floating point argument. Thus, K = SQRT (2.0) will not put 1.4142 into location K. The value returned by INT must be assigned to an integer location; J = INT(Z) is correct, but B = INT(Z) is not. Arguments to the functions can be:

(i) numerical values, e.g. 4.6;

(ii) names of locations or array elements;

(iii) arithmetic expressions which can incorporate (i) and (ii) and even other functions.

For example:

(i) S = SQRT(4.6)

(ii) S = SQRT(V)
S = SQRT(X(I)) (provided location I holds an appropriate integer value)

(iii) S = SQRT((6.7*ABS(X) + FLOAT(J))*72.3)

Table 3.15

```
C
C THIS PROGRAM WILL CONVERT DEGREES FAHRENHEIT TO DEGREES CENTIGRADE.
C THE DATA DECK MUST BE IN THE ORDER:
C [1] ONE CARD SPECIFYING N, THE NUMBER OF DATA CARDS TO FOLLOW.
C         N IS READ IN (I4) FORMAT. I. E.  IN COLS. 1-4 OF THE DATA CARD.
C [2] TO [N+1]: EACH OF THESE DATA CARDS CONTAINS ONE FAHRENHEIT VALUE
C         PUNCHED IN (F10. 0) FORMAT, IE.  IF NO DECIMAL POINT IS PUNCHED
C         IT IS ASSUMED TO FOLLOW COLUMN 10
C
      CONST=5. 0/9. 0
      READ(5, 1) N
    1 FORMAT(I4)
      WRITE(6, 2)
    2 FORMAT(31H1ITEM    FAHRENHEIT    CENTIGRADE)
      DO 4 I=1, N
      READ(5, 3) F
    3 FORMAT(F10. 0)
      C=(F-32. 0)*CONST
    4 WRITE(6, 5) I, F, C
    5 FORMAT(1H , I3, 2F12. 2)
      STOP
      END
```

Example Program 10

The following program (see Table 3.15) will convert degrees Fahrenheit to Centrigrade using the relationship

$$C = (F - 32.0) \times (5/9)$$

H formats will be used to provide headings in the output. A suitable data deck for this program would be:

```
0005
212.0
100.0
 76.5
 32.0
  0.0
```

Remember that if we do not punch the decimal point it is assumed to follow column 10, since we are using (F10.0) as our input format. If we had punched 212 instead of 212.0 then the computer would have interpreted the value as 212ΔΔΔΔΔΔΔ = 2,120,000,000.0 which is not what we intended. We could quite legitimately use the following data deck and get the same results:

```
ΔΔΔ5
ΔΔΔΔΔΔΔ212
ΔΔΔΔΔΔΔ1ΔΔ
ΔΔΔΔΔΔ76.5
ΔΔΔΔΔΔΔΔ32
ΔΔΔΔΔΔΔΔΔΔ
```

This takes advantage of the fact that a blank ('Δ') in an input field is read as a zero. The output from this run of the program is listed below (Table 3.16).

Table 3.16 Sample Output from Example Program 10

ITEM	FAHRENHEIT	CENTIGRADE
1	212.00	100.00
2	100.00	37.78
3	76.50	24.72
4	32.00	0.00
5	0.00	-17.78

Example Program 11

Write a computer program to calculate the product-moment correlation coefficient between two variables X and Y.

Before writing the program let us examine the alternative strategies open to us. There are several ways of computing Pearson's r, for example[1]:

$$r_{xy} = \frac{(x_i - \bar{x})(y_i - \bar{y})}{\{\Sigma(x_i - \bar{x})^2 \Sigma(y_i - \bar{y})^2\}^{\frac{1}{2}}} \qquad (1)$$

$$r_{xy} = \frac{x_i\, y_i - n\bar{x}\bar{y}}{\{(\Sigma x_i^2 - n\bar{x}^2)(\Sigma y^2 - n\bar{y}^2)\}^{\frac{1}{2}}} \qquad (2)$$

$$r_{xy} = \frac{\Sigma x_i y_i - (\Sigma x_i \Sigma y_i)/n}{\{(\Sigma x_i^2 - (\Sigma x_i)^2/n)(\Sigma y_i^2 - (\Sigma y_i)^2/n)\}^{\frac{1}{2}}} \qquad (3)$$

The summation sign (Σ) was explained on page 34. The symbol $(x)^{\frac{1}{2}}$ means: the square root of x, or whatever arithmetic expression lies in the brackets.

Which of the three formulae should be used? Intuitively we would suspect that all three would give the same answer, but our intuition would be reliable only in a perfect world in which calculations were carried out to a large number of decimal places. The basic difficulty is due to the fact that computers are poor at subtracting nearly equal quantities. This rules out Formulae (2) and (3) because the numerators (top lines) in both cases involve subtracting quantities which may, on some occasion, be nearly equal. Despite this Formula (3) is very widely used. It has been shown by several authors that the results of using Formulae (2) and (3) can be very inaccurate, because (i) the computer word (store location) can hold only an approximation to the actual value of the number, (ii) all arithmetical operations are carried out to a fixed number of significant figures, and (iii) the subtraction of nearly equal quantities often results in erroneous answers. Unfortunately, the computer cannot tell that it is going wrong, so that no indication is given on the printout.

It would be unlikely that the results of

[1] Summation is from 1 to n, where n is the number of measurements per variable.

using any of Formulae (1), (2) and (3) would differ significantly when small sets of data are involved, but in trend surface analysis and multiple regression the problem of finding a good numerical algorithm is a serious one that has been largely ignored by geographers.

Formula (1) will therefore be used because it is less likely to result in error. Before trying to write a program based on this formula it would be wise to examine the formula in more detail, and work out a systematic approach to programming it.

(1) set up floating point arrays X and Y of sufficient size to hold the observations on each of two variables;

(2) read the (integral) number of measurements on each variable. Store this in the location called N;

(3) read alphanumeric identifications of variables X and Y and the job name;

(4) read in N measurements into arrays X and Y using (10F8.0) format, i.e. 5 pairs of (X,Y) values are entered per card, with an 8 column field allocated to each value. The decimal point follows the 8th column of each field unless explicitly punched;

(5) compute the means of X and Y;

(6) set the contents of the following locations to zero:

TOPLIN, SSX and SSY;

(7) accumulate the following sums:

$$\begin{aligned} \text{TOPLIN} &= (\text{X(I)} - \text{XBAR})*(\text{Y(I)} - \text{YBAR}) \\ \text{SSX} &= (\text{X(I)} - \text{XBAR})^2 \\ \text{SSY} &= (\text{Y(I)} - \text{YBAR})^2 \end{aligned}$$

summing from I = 1 to I = N;

(8) the correlation coefficient R is now found from

R = TOPLIN/SQRT(SSX*SSY);

(9) output relevant information.

Steps 1, 2 and 4 to 7 should be fairly straightforward as they involve ideas covered in Sections 3.2–3.5. Steps 3 and 9 involve the use of A and H formats, while a library function (SQRT) is used at Step 8. At Step 3, 16 characters (4 locations, each holding 4 characters) are read into arrays XNAME and YNAME, while 80 characters are stored in the array TITLE. It may be confusing at first to find that the same format specification (20A4) is used to place 4 characters into each of the 4 constituent words of XNAME and YNAME and also into the 20 constituent words of TITLE. If you refer to page 40 you will extrapolate from this that the number 20 specifies the maximum number of A4 groups to be read from one data card. Since a card can hold 80 ÷ 4 = 20 sets of 4 characters, (20A4) is as large a specification as possible.

Step 8 is a straightforward use of the function SQRT. At Step 9 the following points are made:

(a) A and H formats can be mixed (formats labelled 6 and 7);

(b) A, H and F formats can be mixed (format labelled 10);

(c) the slash character (/) can be used to print sections of a single format statement on separate lines (formats labelled 6, 7 and 8);

(d) the slash character (/) can be placed at the end of a format specification, as is done in the format labelled 8. Each separate '/' causes a line-feed on the printer; thus, two blank lines will follow the 'Y' at the end of Format 8.

EXERCISES

(24) Use Example Program 11 to find the correlation between the following variables:

	1	2	3	4	5
X (Basin Area)	4.08	4.53	4.92	6.18	6.22
Y (Total stream length)	9.40	10.85	10.69	10.97	11.94

	6	7	8	9	10
X (Basin Area)	6.54	7.55	8.92	10.16	12.50
Y (Total stream length)	18.69	17.21	18.14	20.14	20.68

(Doornkamp and King, 1971, p. 46)

(Hint: the data deck begins with the following cards:

	Column Number
	1 16
data card 1:	0010
data card 2:	BASIN AREA
data card 3:	TOT LNTH STREAMS
	....
	....
	..................).

(25) If the 10 pairs of measurements given in Exercise (1) represent a random sample drawn from a bivariate-normal population then the t test can be used to test the null hypothesis that the population correlation coefficient ρ is zero (H_0: $\rho = 0$) against the alternative hypothesis that it is either positive or negative (H_A: $\rho > 0$ or $\rho < 0$). (N.B. ρ is the Greek rho—it is used to represent the population correlation coefficient. The sample estimator is usually written r, but it should be written $\hat{\rho}$ to conform to normal statistical practice.)

The t-value used to test H_0 is:

$$t = \frac{r\sqrt{n-2}}{\sqrt{1-r^2}}$$

with $(n-2)$ degrees of freedom. H_0 is rejected if the calculated t exceeds the tabled t at a given significance level. Extend Example Program 11 (Table 3.17) so that t and the associated degrees of freedom are printed.

(Hint: the SQRT function requires floating point and not integer arguments.)

(26) Write a program to read in and print out the names and heights of the members of your class.

Determine:

(a) the mean height of the class;

(b) the mean heights of the males and females separately.

(Hints: record height in centimetres or inches. Use the following program strategy:

1. Declare one array, NAME, of dimension 4. Read a job title into this array using (4A4) format, and print it out prior to step 9.

Table 3.17

```
C STEP 1
C
      DIMENSION X(100),Y(100),TITLE(20),XNAME(4),YNAME(4)
C
C STEP 2
C
      READ(5,1) N
    1 FORMAT(I3)
C
C STEP 3
C
      READ(5,2) (XNAME(I),I=1,4)
      READ(5,2) (YNAME(I),I=1,4)
      READ(5,2) (TITLE(I),I=1,20)
    2 FORMAT(20A4)
C
C STEP 4
C
      READ(5,3) (X(I),Y(I),I=1,N)
    3 FORMAT(10F8.0)
```

Table 3.17 (continued)

```
C
C STEP 5
C
      XBAR=0.0
      YBAR=0.0
      DO 4 I=1,N
      XBAR=XBAR+X(I)
    4 YBAR=YBAR+Y(I)
      FN=FLOAT(N)
      XBAR=XBAR/FN
      YBAR=YBAR/FN
C
C STEP 6
C
      TOPLIN=0.0
      SSX=0.0
      SSY=0.0
C
C STEP 7
C
      DO 5 I=1,N
      V=X(I)-XBAR
      W=Y(I)-YBAR
      TOPLIN=TOPLIN+(V*W)
      SSX=SSX+(V*V)
    5 SSY=SSY+(W*W)
C
C STEP 8
C
      R=TOPLIN/SQRT(SSX*SSY)
C
C STEP 9
C
      WRITE(6,6) (TITLE(I),I=1,20)
    6 FORMAT(32H1CORRELATION COEFFICIENT PROGRAM/10HOJOB NAME ,20A4)
      WRITE(6,7) (XNAME(I),I=1,4),(YNAME(I),I=1,4)
    7 FORMAT(24HOVARIABLE NAMES ARE X = ,4A4,8HAND Y = ,4A4)
      WRITE(6,8)
    8 FORMAT(22HOLISTING OF INPUT DATA/20HO         X          Y//)
      WRITE(6,9) (I, X(I), Y(I), I=1,N)
    9 FORMAT(1H ,I3,2F10.4)
      WRITE(6,10) (XNAME(I),I=1,4),(YNAME(I),I=1,4), R
   10 FORMAT(21HOCORRELATION BETWEEN ,4A4,5H AND ,4A4,3H IS,F10.4)
      STOP
      END
```

2. Read NO1, the number of males and NO2, the number of females;
3. Set HTMEN = 0.0;
4. For I = I to I = NO1 (DO loop)
   (a) read the subject's name in (4A4) format and his height in (F4.0) format, using two data cards;
   (b) print out this information;
   (c) add the subject's height to the current contents of location HTMEN, replacing the previous contents;
5. Divide the contents of location HTMEN by the floating point equivalent of the value stored in NO1.[1] This is the mean height of the male group. Store the result in XMEN, and print out its value;
6. Set HTWOM = 0.0;
7. Repeat Step 4 using a DO loop from I = 1 to I = NO2, and replacing HTMEN with HTWOM;
8. Divide the contents of HTWOM by the floating point equivalent[1] of the value in NO2. This is the mean height of the female group. Store the result in XFEM, and print its value;
9. The average class height is $\frac{1}{2}$(XFEM + XMEN). Store the result and print it out.

(27) Using Example Program 10 (Table 3.15) as a guide, write a program to convert miles to kilometres, using the fact that 1 mile = 1.6093 kilometres.

[1] Function FLOAT (INTEGER) gives the floating point equivalent of an integer argument.

Summary:

1. Alphanumeric data (characters + numerals + special symbols such as * ( ) =) can be input and output using A format specification (pages 40–41).
2. Fixed strings of alphanumeric data can be output using H format descriptions (pages 41–42).
3. No H or A format should specify that more than 120[1] characters be output on any one line.
4. An A format on input should not specify that more than 80 characters be read from any one data card.
5. A formats are of the form: nAm where n is the number of locations to be filled and m is the number of characters stored in each location. A maximum of 4 characters can be assigned to each location[2].
6. H formats are of the form: nHcxxx .... where n is the number of characters following the H, c is a carriage control character, and xxx .... are the alphanumeric characters to be output.
7. A slash (/) can be used within a format specification to cause printing to resume on the next line. Multiple slashes can also be used—one new line is output for each slash.
8. Alphanumeric data can be read into variables or arrays of type floating point or type integer, provided that the two are not intermixed within one program.
9. Format specification statements (or any other Fortran statements) that are too long to fit on one card can be continued on the next card provided that any character other than the digit zero (0) is punched in column 6 of each continuation card.
10. Comments can be punched within the program by entering a C in column 1 of the card.
11. Columns 73–80 can be used for program card identification.
12. Library functions can be used to calculate the following values: square root (SQRT), log base e (ALOG), log base 10 (ALOG10), e to the power of (EXP), sine (SIN), cosine (COS), nearest integer to (INT), absolute value of (ABS) and floating point equivalent of (FLOAT) (p. 42).
13. Function arguments must be of the specified type (p. 43).
14. The value returned by a function should be assigned to a location of the correct type (p. 44).

[1] Or 132 or 160, depending on the lineprinter.
[2] Some computers can fit 6, 8 or 10 characters per location but all can accommodate at least 4.

CHAPTER 4

Applications

4.1 Introduction

By this stage you should know enough Fortran to write useful programs of your own, and to be able to use some of the published Fortran programs which are mentioned in Appendix B. In this chapter a series of more advanced programs will be presented and discussed. These programs are intended not only to be of practical use, but also to show how a problem should be approached. In fact, the planning stage is often more demanding than the actual programming task—it is relatively easy to program a problem once you have worked out how to approach it. The first program, Example Program 12, deals with the analysis of the properties of spatial distributions of points using nearest-neighbour methods. Example Program 13 computes the Spearman rank correlation coefficient rho, which is easier to understand than the Pearson coefficient r which was programmed in Chapter 3 (Example Program 11). However, the programming strategy involved in finding rho is slightly more involved than that used in obtaining the value of r. The third program (Example Program 14) introduces linear programming techniques. (The word 'programming' in 'linear programming' does not refer to computer programming—it should be interpreted in the sense of scheduling.) The fourth program in this chapter involves the simulation of stream networks (Example Program 15). Computer simulation is a method of experimenting, using the computer as a sort of laboratory. It allows a user to test hypotheses in the absence of observational data. The last example program computes the linear regression function and associated statistics.

Some background reading will be necessary if these programs are to be understood fully (refer to Appendix E).

4.2 Nearest-Neighbour Method

The properties of spatial point patterns are of great interest to geographers, provoking questions such as: are these points randomly distributed over space? are they more clustered than random or more uniform than random? Figure 4.1 shows random, clustered and uniform point patterns. The points could represent settlements, or rock outcrops, or any other geographical phenomenon. One method of answering the questions posed above is provided by nearest-neighbour analysis. The theory underlying this technique has been covered adequately elsewhere, and readers should refer to Ebdon (1976) for details. Initially, as the name suggests, the method involves the determination of the nearest neighbour of each point. The nearest neighbour of a point is the point that is closest to it. Next, the distance from a point

Figure 4.1A. Random point pattern

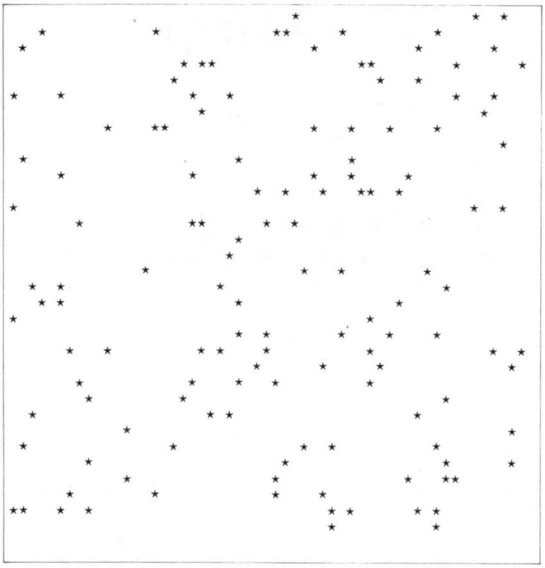

Figure 4.1B. Clustered point pattern

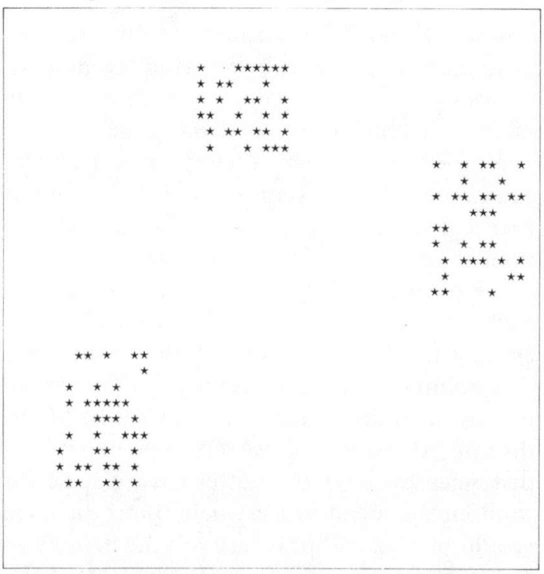

to its nearest neighbour is calculated using Pythagoras's theorem. If the map co-ordinates of a pair of points P and Q with respect to a rectangular grid are known, then the distance PQ is calculated as follows, assuming the co-ordinates of point P to be (6,2) and of Q to be (3,4):

$$PQ = {}^{+}\sqrt{(6-3)^2 + (2-4)^2}$$
$$= {}^{+}\sqrt{13} \quad = \quad 3.604$$

Figure 4.1C. Uniform point pattern

(N.B. The symbol ${}^{+}\sqrt{x}$ means: the positive square root of the value x. For example, ${}^{+}\sqrt{4} = 2$.) These calculations are shown in Figure 4.2. In Figure 4.2 (see p. 52) the required distance is the length of hypotenuse of the right-angled triangle PQR. Pythagoras's theorem states that $PQ^2 = QR^2 + RP^2$. The distance RP is given by the difference between the first co-ordinates of points P and Q, i.e. $6 - 3 = 3$ while $RP = 4 - 2 = 2$. It therefore follows that

$$PQ^2 = 3^2 + 2^2 = 9 + 4 = 13,$$

hence

$$PQ = +\sqrt{13} = 3.604.$$

This gives us the following computing algorithm for determining PQ:

Step 1: place the co-ordinates on axis x1 of the points P and Q into the first two storage locations of the floating point array HORIZ. Similarly, place the co-ordinates on axis x2 into the floating point array VERT.

Step 2: the distance PR is found from the expression HORIZ(2) – HORIZ(1). (Note: the sign of the resulting

Figure 4.2

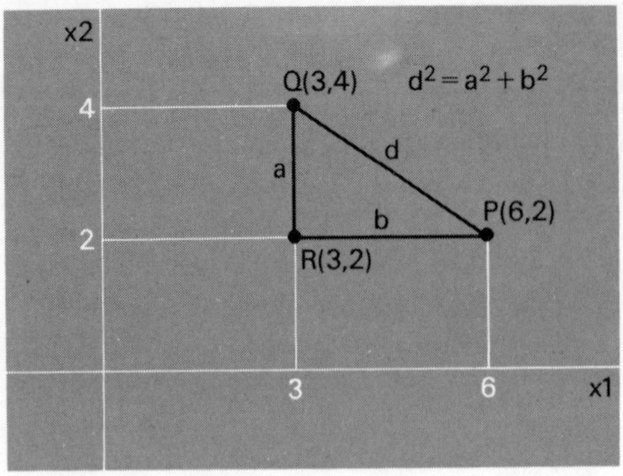

Table 4.1

```
  DIMENSION HORIZ(2),VERT(2)
  HORIZ(1)=6.0
  HORIZ(2)=3.0
  VERT(1)=2.0
  VERT(2)=4.0
  PR=HORIZ(2)-HORIZ(1)
  QR=VERT(2)-VERT(1)
  PQSQD=PR*PR + QR*QR
  PQ=SQRT(PQSQD)
  WRITE(6,1) PQ
1 FORMAT(6HOPQ = ,F10.4)
  STOP
  END
```

number may be positive or negative, but squaring achieves a positive result even when the value to be squared is negative.)

Step 3: the distance QR can be found from VERT(2) – VERT(1).

Step 4: PQ^2 is obtained by squaring the values obtained at Step 3 and at Step 2, and adding them.

Step 5: PQ, the required distance, is the square root of the value obtained at Step 4. Note that the Fortran library function SQRT always returns the positive square root.

This algorithm is coded in Fortran in Table 4.1.

The algorithm can be generalized to deal with all the $n(n-1)/2$ distances between every pair out of a set of n points, but the basic rule remains: the distance between any pair of points is the positive square root of the sum of the squared differences between the co-ordinates of the two points on axes x1 and x2. This can be stated more concisely as:

$$d_{ij} = +\sqrt{\{(x1_i - x1_j)^2 + (x2_i - x2_j)^2\}}$$

In this equation, d_{ij} is the distance between point i and point j, and $(x1_i, x2_i)$, $(x1_j, x2_j)$ are the co-ordinates of the two points on axes x1 and x2. This statement is written in Fortran as

DIST = SQRT ((X1(I) – X1(J))**2 + (X2(I) – ↔
↔ X2(J))**2 (1)

where X1 and X2 are arrays holding the co-ordinates of the points on axes x1 and x2 respectively. The symbol ** means: to the power of. Thus 6.0**3 is 216.0.

Nearest neighbour analysis requires the evaluation of the distances from a given point to the remaining $n-1$ points in a set of n points, and the selection of the shortest of these distances. This process is repeated for each of the n points in turn, so that, on completion of the process, the n nearest-neighbour distances are known. The mean nearest neighbour distance can then be found dividing the sum of these n nearest neighbour distances by n. If the pattern were spatially random the mean nearest-neighbour distance would be $1/(2 \times \sqrt{p})$ where p is the density of points per unit area. (If A is the area of the region of interest, in square units, and n the number of points, then $p = A \div n$.) The ratio between the calculated mean nearest-neighbour distance and the value expected if the pattern were spatially random is R, the nearest-neighbour statistic. R lies between 0 and 2.15. An R-value of 0 indicates that the n points lie on top of each other, while an R-value of 1.0 shows that the pattern is spatially random. An R-value of 2.15 is produced by a regular or uniform point pattern.

To compute R for a set of n points, the following algorithm is suggested:

Step 1: Set up two arrays of floating point values, X1 and X2.

Step 2: Read n, the number of points and store it in location N. Read the n sets of (X1,X2) co-ordinates of these points and store them in the floating point arrays X1 and X2. Set the contents of location DMEAN to zero, and place the floating point equivalent of the value n into location FN.

Step 3: Compute the distance from the first point to each of the other points in turn. Add the smallest of these distances to the contents of location DMEAN. Repeat Step 3 for all of the n points, so that eventually DMEAN will contain the sum of the n nearest-neighbour distances.

Step 4: Divide the contents of DMEAN by the floating point equivalent of the contents of N, stored in FN. The result is the (calculated) mean nearest-neighbour value.

Step 5: Read into a floating point location called AREA the area of the region being studied. Put in location EXPECT the value of the expected mean nearest-neighbour distance, found from EXPECT = AREA/FN.

Step 6: Compute the ratio of the contents of DMEAN and the contents of EXPECT, and place the resulting value in location R.

Step 7: Print the contents of R and terminate the program.

Table 4.2

```
      DMEAN=0.0
C N IS NUMBER OF POINTS
      DO 2 I=1,N
      DMIN=1000000.0
C I IS THE INDEX OF THE CURRENT POINT.  NOW COMPUTE DISTANCES FROM
C POINT I TO ALL OTHER POINTS.
      DO 3 J=1,N
C EXCLUDE DISTANCE FROM POINT I TO ITSELF.
      IF(I.EQ.J) GOTO 3
C NOW COMPUTE DISTANCE FROM CURRENT POINT (I) TO POINT J USING
C FORMULA (1).  THE SQUARE OF THE DISTANCE IS COMPUTED BECAUSE WE ARE
C ONLY INTERESTED IN FINDING THE SMALLEST DISTANCE.  THUS, THE SQUARE
C ROOT OPERATION IS CARRIED OUT ON THE SMALLEST (SQUARED) DISTANCE
C FOUND IN LOOP 3.
      DIST=(X1(I)-X1(J))**2+(X2(I)-X2(J))**2
      IF(DIST.LT.DMIN) DMIN=DIST
C IF COMPUTED SQUARED DISTANCE IS LESS THAN THE VALUE CURRENTLY HELD
C IN LOCATION DMIN THEN OVERWRITE THE CONTENTS OF DMIN WITH THE CURRENT
C SQUARED DISTANCE.
    3 CONTINUE
C A CONTINUE STATEMENT IS A DUMMY STATEMENT USED TO END A DO-LOOP
C WHEN A GOTO STATEMENT CAUSES CONTROL TO BE TRANSFERRED TO THE LAST
C STATEMENT OF THE DO-LOOP.  IT CAN BE INTERPRETED AS AN INSTRUCTION TO
C DO NOTHING.
    2 DMEAN=DMEAN+SQRT(DMIN)
```

Steps 1 and 2, and 4 to 7 present no problems. Step 3 involves the comparison of the magnitudes of all the distances from a point to the other (n – 1) points. In Example Program 8 (Table 3.10) a conditional (IF) statement was used to check whether Newton's square-root algorithm had converged. The form of the IF statement is:

IF (condition) executable statement

The use of an IF statement at Step 3 of the algorithm can be illustrated as follows. Set the contents of the floating-point location DMIN to a very large positive floating point value, say 1000000.0. Calculate the distance from the current point to each of the other (n – 1) points in turn, using Formula (1), above. Every time a distance is calculated it is stored in the floating point location DIST, and the contents of DIST and DMIN are

Table 4.3

```
C               EXAMPLE PROGRAM 12.
C STEP 1
      DIMENSION X1(100),X2(100)
C STEP 2
      READ(5,100) N
  100 FORMAT(I3)
      WRITE(6,500) N
  500 FORMAT(16HOCOORDINATES FOR,I4,8H  POINTS//)
      DO 1 I=1,N
      READ(5,200) X1(I), X2(I)
    1 WRITE(6,400) I, X1(I), X2(I)
  400 FORMAT(1H ,I4,2F10.4)
  200 FORMAT(2F10.0)
      FN=N
      DMEAN=0.0
C STEP 3
      DO 2 I=1,N
      DMIN=1000000.0
      DO 3 J=1,N
      IF(I.EQ.J) GOTO 3
      DIST=(X1(I)-X1(J))**2+(X2(I)-X2(J))**2
      IF(DIST.LT.DMIN) DMIN=DIST
    3 CONTINUE
    2 DMEAN=DMEAN+SQRT(DMIN)
      DMEAN=DMEAN+SQRT(DIST)
C STEP 4
      DMEAN=DMEAN/FN
C STEP 5
      READ(5,200) AREA
      EXPECT=AREA/FN
C STEP 6
      R=DMEAN/EXPECT
C STEP 7
      WRITE(6,300) R
  300 FORMAT(31HONEAREST NEIGHBOUR STATISTIC = ,F8.4)
      STOP
      END
```

compared, using an IF statement. If the contents of DIST are smaller in magnitude than the current contents of DMIN then the contents of DMIN are overwritten by the contents of DIST. The Fortran instructions for Step 3 are shown in Table 4.2 (see p. 53). Example Program 12 is listed in Table 4.3.

EXERCISE

(28) Assume that the distances from the current point to each of the other 5 points in a set of 6 points are 6.3, 8.2, 7.4, 3.7 and 4.1. Use the following layout to show that the contents of DMIN are eventually equal to 3.7.

DMIN	1000000.0	6.3	6.3
DIST	undefined	6.3	8.2
CYCLE NO	undefined	1	2

Example Program 12 extends some ideas discussed earlier. First of all, Step 3 involves the use of two DO loops, one 'inside' the other. This causes the loop DO 3 J = 1,N to be carried out for each of the N values J = 1, 2, ..., N for every I value. Now I takes the values 1 to N also, so the inner loop is executed N^2 times. A good analogy can be made between this use of nested DO loops and the hands of a clock. The minute hand represents the inner loop (DO 3 J = 1,N), the hour hand the outer loop (with counter I). For each full revolution of the minute hand the hour hand moves on one hour, so in 12 hours the hour hand has moved 1 revolution and the minute hand has moved 12. There could be a seconds hand as well—this would correspond to a triple nested DO loop. The most widespread fault in the use of nested DO loops is the failure to give each loop a separate loop index. We have used I as the index for the outer loop and J for the inner loop. The following Fortran statements would produce an unsatisfactory result because the same index, I, is used in both DO loops simultaneously:

```
┌──────DO 5 I = 1,N
│ ┌────DO 4 I = 1,N
│ │    ...........
│ └4   ...........
└──5   ...........
```

The second fault is caused by failure to 'nest' the loops properly. Thus (a) is correct, (b) incorrect:

```
(a)
┌──────DO 6 I = 1,N
│ ┌────DO 5 J = 1,2
│ │    ...........
│ └5   ...........
└──6   ...........

(b)
  ┌────DO 6 I = 1,N
┌─┼────DO 5 J = 1,2
│ │    ...........
│ └6   ...........
└──5   ...........
```

In (b) the DO loops overlap. This is illegal, and will result in compile-time errors (Appendix C).

EXERCISES

(29) How will the data cards be punched for Example Program 12? Prepare a sample data listing for n = 10 points.

(30) Explain why it is legal to use the format labelled 200 in Example Program 12 to input a single value, stored in AREA, in Step 5, when the format specification is (5F10.0).

(31) Use Example Program 12 to find the nearest-neighbour statistic for the following 10 points:

Point No.	1	2	3	4	5
X1 co-ord.	2.7	6.1	3.7	2.4	5.9
X2 co-ord.	6.8	2.9	0.0	1.6	4.7

Point No.	6	7	8	9	10
X1 co-ord.	8.1	6.2	7.3	8.1	5.0
X2 co-ord.	1.6	6.1	5.2	2.9	2.7

Assume the area of the region to be 37 square units. Plot the points on a graph. Compare the results of a purely visual analysis with the nearest-neighbour results.

(32) The power of a given number can be computed using the operator '**' which means 'to the power of'. For example,

A = B**3

will cause the cube of the value stored in location B to overwrite the contents of location A. Hence, powers of numbers or arithmetic expressions need not be computed by repeated multiplication, though it is slightly more efficient to write Y = X*X rather than Y = X**2.

Notes:

(i) The ** operator is carried out before * and /, which in turn have precedence over + and −.

(ii) An integer value may not be raised to the power of a floating point value, i.e.

I**FL

is *illegal*. However, a floating point value can be raised to the power of an integer value or a floating point value; both

FL**I and FL**FL

are legal. (I is an integer location, FL a floating point location.)

(iii) A negative value cannot be raised to the power of a floating point exponent, so (−6.3)**2.0 is illegal.

(iv) If an expression of the type $(x^y)^z$ is to be evaluated it must be written as

(X**Y)**Z

and not as X**Y**Z.

(v) If both X and Y are zero in the instruction S = X**Y, failure will occur.

(vi) Beware of raising integers to a negative power. Because of rule (ii) the exponent must always be an integer if the base is an integer. However, the result is also of type integer, so that 3**(−1) would give the value 0 ($3^{-1} = \frac{1}{3}$, but $\frac{1}{3}$ is not an integer; the nearest integer is 0). However, 3.0**(−1) would correctly return the value 0.333.

L. J. King (1969, pages 92 and 93) describes the measure known as Bachi's standard distance statistic. Given n points with coordinates $(x1_i, x2_i)$ with respect to the orthogonal axes x1 and x2, the distance d_{ij} from point i to point j is:

$$d_{ij} = [(x1_i - x1_j)^2 + (x2_i - x2_j)^2]^{\frac{1}{2}}$$

The 'mean quadratic distance' is D, where

$$D = [(\sum_{i=1}^{n} \sum_{j=1}^{n} d^2_{ij})/n^2]^{\frac{1}{2}}$$

The mean centre of the point pattern is then given by the coordinates $\overline{x1}$ and $\overline{x2}$:

$$\overline{x1} = (\sum_{i=1}^{n} x1_i)/n, \quad \text{and} \quad \overline{x2} = (\sum_{i=1}^{n} x2_i)/n$$

If d_{ic} is the distance from point i to the mean centre, then the standard distance is defined as:

$$d = \left[\frac{\sum_{i=1}^{n} d^2_{ic}}{n} \right]^{\frac{1}{2}}$$

Write a Fortran program to compute the mean quadratic distance, the mean centre and the standard distance for up to 100 points. You can use the following points as test data:

Crossens (3.378,4.206);
Rufford (3.456,4.164),
Southport (3.371,4.208);
Worthington (3.580,4.102),
Wigan (3.557,4.066),
Billinge (3.525,4.030);
Mawdesley (3.493,4.150);
Dark Lane (3.428,4.088);
Lathom (3.458,4.086);
Halewood (3.460,3.859);
Speke (3.415,3.839);
Garston (3.395,3.853);
Knotty Ash (3.411,3.915);
Kirkby (3.399,3.980);
Melling (3.393,4.009);
Bickerstaffe (3.419,4.036);

Springfield (3.388,4.046);
Ainsdale (3.318,4.112);
Hightown (3.294,4.044);
Liverpool (3.361,3.950).

The data are the Ordnance Survey grid references of 20 rainfall recording stations in South-west Lancashire.

(33) If the points in a point pattern form a sample drawn from a larger population, and if the population distribution of nearest-neighbour distances can be assumed to be normal, it is possible to test the null hypothesis that the nearest-neighbour statistic is in fact 1 in the population (i.e. the sample value deviates from 1 by chance alone).

The test involves the calculation of the standard error of the observed mean nearest-neighbour distance, given by

$$SE\bar{d}_o = 0.26136/n \times p$$

where $SE\bar{d}_o$ is the standard error, n the number of points and p the density of points. The statistic c can be calculated from

$$c = (\bar{d}_o - \bar{d}_e)/SE\bar{d}_o$$

where $\bar{d}_o$ is the observed mean nearest-neighbour distance and $\bar{d}_e$ the expected mean nearest-neighbour distance under the assumption of randomness. The statistic c is normally distributed with a mean of zero and a variance of 1, so that the calculated value of c can be looked up in a table of the standard normal distribution. (For example, a value of c greater than $\pm$ 1.65 will occur in only 5% of a large number of trials when the null hypothesis is true, and a value of c greater than 2.33 will occur in only 1% of such trials.)

Extend Example Program 12 to provide for the calculation and printout of the value of c.

4.3 Spearman's Rank Correlation Coefficient

This program is designed to calculate the Spearman rank correlation coefficient ρ ('rho') either from raw data or from ranked data. Given a set of rank-order data measured for n individuals on two variables, x and y, rho is easily calculated. For example, if the observations on 6 individuals are ranked as follows:

		Individual					
Variable	x	1	3	2	4	6	5
	y	2	1	6	3	4	5

where '1' indicates 'largest value', '2' indicates 'second largest', and so on, then

$$\rho = 1.0 - \frac{6\sum_{i=1}^{n} d_i^2}{n^3 - n} \qquad (2)$$

In this formula, d_i^2 is the squared difference between the corresponding (ranked) values of x_i and y_i. For the data given above, the d_i^2 are as follows:

d_1^2	d_2^2	d_3^2	d_4^2	d_5^2	d_6^2
1	4	16	1	4	0

and $\sum_{i=1}^{6} d_i^2 = 26$. The value of rho is:

$$\rho = 1 - \frac{6 \times 26}{6 \times 6 \times 6 - 6} = 1 - \frac{156}{210} = 0.2572.$$

Thus far, the calculations are simple.

EXERCISE

(34) Write a program to read in ranked values, such as the x and y values given above, and calculate (i) and d_i; (ii) the d_i^2 and (iii) rho. Print out (a) the rank values and (b) rho.

Given a set of scores (measurements) on x and y the ranks have to be derived, if they are not already available. This is not quite as simple as it first appears. If the actual measurements on variable x, rather than the ranks given above, are:

x: 100 60 72 55 10 20

then, simply by examining the values visually, one could pick out 100 as the largest value, 72 as the second largest value, and so on. This can only be done if the whole set of measurements can be seen simultaneously. If the measurements are numerous, then a different approach must be used. For example, if we are asked to rank the following values:

6, 3, 10, 4, 8, 7, 9, 2, 11, 14, 21, 16, 3.5, 1.6, 17, 20, 8.1, 19, 32, 15, 17.2, 16.8, 30.9, 16.75, 3.9, 1, 18, 2.9, 6.4

then the following method could be used. Firstly, compare the first two values and write down the larger, followed by the value of its position in the list, e.g. 1 or 2. Next, compare the number just written down with the third number in the list; if the 3rd number is larger, write it down together with its position in the list, 3, and delete the previous two numbers. Now compare the fourth number in the list with that written down, and proceed in a similar manner until all the numbers in the list have been compared. The last pair of numbers written down will be the highest number in the list and its numerical position. For example, if this procedure was followed with the list of numbers given above, we would write down the following sequence:

~~6 1~~ ~~10 3~~ ~~11 9~~ ~~14 10~~

~~21 11~~ 32 19

Thus, the highest value in the sequence is 32 and it comes in the 19th position. If we delete the value 32 from the list and repeat the process from the beginning we would get the second highest value, and so on.

This apparently cumbersome method is the one we will use in Example Program 13 (though, inevitably, faster methods have been developed—in fact, finding the fastest sorting algorithm seems to be a challenge to all computer programmers). The algorithm is:

Step 1: Set up two floating point arrays, X and RANKX, to hold the observations and the ranks, respectively. Read the value of n, the number of observations. Store it in location N. Put the n observations into the first n locations of the array X. Set an integer location J to the value 1.

Step 2: Set a real location A to a number that is less than any of the observations stored in X. A = –1000000.0 will be less than most observations we are likely to meet in practice, so we can arbitrarily use this value. Store the integer value 1 in location I.

Step 3: Compare the contents of X(I) with those of A. If X(I) holds a number that is smaller than, or equal to, that stored in A then go to Step 4. Otherwise:
(3a) replace the contents of A by the contents of X(I) and set the contents of an integer location K to the value of the contents of I.

Step 4: Increase the value of the contents of I by 1. If the value stored in I is greater than the value stored in N go to Step 5, otherwise go to Step 3.

Step 5: Put the floating point equivalent of the integer value stored in J into the Kth element of the array RANKX. Increase the value of the contents of location J by 1. If J is greater than N go to Step 6, otherwise set X(K) to –100000000.0 and go to Step 2.

Step 6: The ranks are now stored in the array RANKX.

The algorithm is illustrated by the following worksheet:

Step 1:	X	1.5	0.9	3.7	5.0
	RANKX	?	?	?	?
	N	4			
	J	1			
Step 2:	A	–1000000.0			
	I	1			
Step 3:	Since 1.5 is greater than –1000000.0 then:				
Step 3a:	A	1.5			
	K	1			

Step 4: I 2
Since 2 is not greater than 4, return to Step 3.
Step 3: Since 0.9 is not greater than 1.5 go to Step 4.
Step 4: I 3
Since 3 is not greater than 4 go to Step 3.
Step 3: Since 3.7 is greater than 1.5 then:
Step 3a: A 3.7
K 3
Step 4: I 4
Since 4 is not greater than 4 go to Step 3.
Step 3: Since 5.0 is greater than 3.7 then:
Step 3a: A 5.0
K 4
Step 4: I 5
Since 5 is greater than 4 go to Step 5.
Step 5: RANKX ? ? ? 1
J 2
Since 2 is not greater than 4 then:
X 1.5 0.9 3.7 −100000000.0
Step 2: A −1000000.0
I 2

and so on. Complete the worksheet and show that the contents of RANKX are eventually

3 4 2 1

EXERCISE

(35) Write a Fortran program based on the above algorithm.

One of the difficulties met with in ranking data is that tied (or equal) values sometimes occur. For example, the data might consist of the values of air temperature recorded at midday on each of seven consecutive days. These values could be: 61°, 62°, 70°, 62°, 65°, 69° and 69°. The strategy given earlier in this chapter would, if applied to the temperature data, give the following result: 7 5 1 6 4 2 3 but—since there are two equal 62's and two equal 69's—the result should be: 7 $5\frac{1}{2}$ 1 $5\frac{1}{2}$ 4 $2\frac{1}{2}$ $2\frac{1}{2}$. In other words, the ranks of equal data values are the average of the ranks they would have had if they had all been slightly different from each other. There are efficient methods of dealing with the problem of equal ranks, but the method used earlier is adapted because it is quite easy to understand how it works. The method involves a modification to Steps 1 and 5 of the ranking algorithm as follows:

Step 1(b): Declare an additional (integer) array IWORK with the same dimension as arrays X and RANKX.

Step 5: Go through array X to find if any elements are equal in value to the number stored in location A, which contains the highest value found at Steps 3 and 4. Keep a count of the number of equal values using the integer location NO. Also, note the positions of these equal elements and store the values of their numerical identifiers in the array IWORK. Thus, the highest value would have a numerical identifier given by the contents of location K, so IWORK(1) would be set to the value of the number held in K. The first equal value would have its numerical identifier stored in IWORK(2), and so on. The average of the ranks that these equal elements would have had if they had all been slightly different is $\frac{1}{2}(I + I - 1 + NO)$. Put this value into all the elements of array RANKX that have identifiers equal to those stored in the first NO elements of array IWORK, and put the value − 100000000.0 into corresponding elements of X. Increase the value of J by NO − 1. If J is greater than N, go to Step 6 otherwise go to Step 1.

EXERCISE

(36) Carry out the modified ranking algorithm on the data shown in Table 4.4, which is a

set of annual maximum peak instantaneous discharges recorded for the years 1928 to 1958 inclusive.

Table 4.4. Data Matrix for Exercise 36—Peak Annual Discharges, Seneca Creek, Dawsonville, Maryland, USA (from L. B. Leopold, 'Water', W. H. Freeman & Co., San Francisco, 1974).

Year	*Discharge*	*Year*	*Discharge*
1928	3800	1943	3620
1929	1600	1944	2660
1930	1450	1945	2110
1931	1730	1946	2940
1932	1380	1947	1990
1933	9300	1948	1990
1934	2410	1949	2240
1035	1420	1950	2280
1936	2020	1951	2420
1937	2610	1952	2810
1938	2280	1953	7330
1939	2150	1954	1240
1940	1740	1955	2620
1941	1300	1956	15000
1942	1460	1957	959
		1958	3640

One point of technical detail should be carefully noted. At Step 5 of the modified algorithm a search is made for equal elements in the floating point array X. In Chapter 2 it was pointed out that floating point values are not necessarily stored exactly, unlike integer values. Thus, the value result of the operation 6.0/2.0 might be stored as 2.999999 or 3.000001 in a computer working to seven decimal places. For all practical purposes, all values in the range 2.999999 to 3.000001 are equal. Instead of using the condition '.EQ.' when comparing floating point values it is a better policy to check whether the absolute difference (i.e. regardless of sign) between two floating point quantities is less than a specified amount, which would be 0.000002 in the above example. The statements

```
IF (X.EQ.Y) STOP
IF (Z(K).EQ.0.0) I=I+1
```

are not wrong, but they are badly expressed. Instead, it would be better to use the alternative statements

```
IF (ABS(X-Y).LT.0.000002) STOP
```

and

```
IF (ABS(Z(K)-0.0).LT.0.000002) I=I+1
```

This difficulty does not occur when integer storage is used. The statements

```
IF (I.EQ.5) STOP
```

and

```
IF (I*7.LT.99) I=I/2
```

are perfectly acceptable.

Once the ranked values of the measurements on the two variables are known, the determination of Spearman's rho is straightforward. Equation (2) stated that

$$\rho = 1.0 - \frac{6\sum_{i=}^{n} d_i^2}{n^3 - n}$$

where d_i^2 is the squared difference between corresponding ranks of x and y and Σ is the summation sign.

EXERCISES

(37) Use Example Program 13 to compute the rank correlation coefficients between variables 1 and 2, 1 and 3 and 2 and 3 in the list below. Variable 1 is persons employed in agriculture as a percentage of total em-

Table 4.5

```
C PRGRAM TO COMPUTE SPEARMAN'S RANK CORRELATION COEFFICIENT (RHO).
C EXAMPLE PROGRAM 13
C DATA INPUT CAN BE SCORES OR RANKS
      DIMENSION X(60),Y(60),RANKX(60),RANKY(60),XNAME(4),YNAME(4),
     1IWORK(60)
C IF NUMBER OF INDIVIDUALS(CASES,AREAS) EXCEEDS 60 THEN ADJUST THE
C DIMENSION STATEMENT.
      READ(5,1) N
C N IS THE NUMBER OF INDIVIDUALS
    1 FORMAT(I4)
      READ(5,1) INDIC
C INDIC = 1 THEN PROGRAM READS RAW SCORES AND THEN RANKS THEM;
C IF INDIC = 2 PROGRAM READS IN RANKS
      READ(5,8) (XNAME(I),I=1,4)
      READ(5,8) (YNAME(I),I=1,4)
C XNAME AND YNAME ARE THE ALPHANUMERIC NAMES OF THE X AND Y VARIABLES.
C UP TO 16 CHARACTERS PER NAME, WITH EACH NAME ON A SEPARATE CARD.
    8 FORMAT(20A4)
      IF(INDIC.EQ.2) GOTO 5
    2 READ(5,3) (X(I),I=1,N)
      READ(5,3) (Y(I),I=1,N)
C READ DATA FOR X AND Y IN TURN, USING (8F10.0) FORMAT. START EACH SET
C OF OBSERVATIONS ON A NEW CARD
C CHANGE FORMAT LABELLED 3 IF A DIFFERENT INPUT FORMAT IS REQUIRED.
    3 FORMAT(8F10.0)
C WRITE OUT X,Y DATA TO LINEPRINTER.
      WRITE(6,12)
   12 FORMAT(35H1INPUT DATA - NOTE THESE ARE SCORES//)
      WRITE(6,4) (I,X(I),Y(I),I=1,N)
    4 FORMAT(1H ,I4,2F10.4)
C RANK THE X'S
      I=0
   30 I=I+1
      A=-100000.0
      DO 31 J=1,N
      IF(X(J).LE.A) GOTO 31
      K=J
      A=X(J)
   31 CONTINUE
C A IS NOW THE LARGEST X VALUE
      NO=0
C NOW CHECK FOR EQUAL VALUES OF A
      DO 32 J=1,N
      IF(ABS(X(J)-A).GT.0.0000001) GOTO 32
C COUNT THE NUMBER OF EQUAL VALUES IN *NO*
      NO=NO+1
```

Table 4.5 (continued)

```
C STORE THE IDENTIFIERS OF THE EQUAL VALUES IN *IWORK*
      IWORK(NO)=J
   32 CONTINUE
C CALCULATE THE AVERAGE RANK FOR THE EQUAL CASES
      SUM=FLOAT(I+I+NO-1)/2.0
C PUT THE AVERAGE RANK IN THE APPROPRIATE ELEMENT OF *RANKX*
      DO 33 J=1,NO
      K=IWORK(J)
      RANKX(K)=SUM
C SET THE VALUE IN *X(K)* TO A VERY LOW NEGATIVE VALUE SO THAT IT WILL
C NOT BE PICKED OUT AT SUBSEQUENT CYCLES.
   33 X(K)=-1000000.0
      I=I+NO-1
      IF(I.LT.N) GOTO 30
C NOW RANK THE Y'S IN THE SAME WAY.
      I=0
   40 I=I+1
      A=-100000.0
      DO 41 J=1,N
      IF(Y(J).LE.A) GOTO 41
      K=J
      A=Y(J)
   41 CONTINUE
      NO=0
      DO 42 J=1,N
      IF(ABS(Y(J)-A).GT.0.0000001) GOTO 42
      NO=NO+1
      IWORK(NO)=J
   42 CONTINUE
      SUM=FLOAT(I+I-1+NO)/2.0
      DO 43 J=1,NO
      K=IWORK(J)
      RANKY(K)=SUM
   43 Y(K)=-1000000.0
      I=I+NO-1
      IF(I.LT.N) GOTO 40
C READ IN RANKS IF *INDIC* WAS SET TO 2
      GOTO 10
    5 READ(5,3)(RANKX(I),I=1,N)
      READ(5,3)(RANKY(I),I=1,N)
   10 WRITE(6,13)
C OUTPUT RANKS OF X AND Y TO THE LINEPRINTER.
   13 FORMAT(12H1RANKED DATA//)
      WRITE(6,14) (I, RANKX(I), RANKY(I),I=1,N)
   14 FORMAT(1H ,I4,2F12.4)
C START CALCULATING THE SUM OF THE SQUARED DIFFERENCES BETWEEN THE RANKS
```

Table 4.5 (continued)

```
C OF X AND Y.  PUT THE (ACCUMULATING) TOTAL IN LOCATION *D*
      D=0.0
      DO 6 I=1,N
    6 D=D+(RANKX(I)-RANKY(I))**2
C RHO IS SPEARMAN'S RANK CORRELATION FOR X AND Y.
      RHO=1.0-D*6.0/FLOAT(N**3-N)
C NOTE THE USE OF THE FUNCTION 'FLOAT' TO CONVERT FORM INTEGER TO REAL
C IN THE ASSIGNMENT STATEMENT SETTING THE VALUE OF RHO.
C NOW OUTPUT THE VALUE OF RHO AND THE NAMES OF THE VARIABLES.
      WRITE(6,7) (XNAME(I),I=1,4),(YNAME(I),I=1,4), RHO
    7 FORMAT(26H0RANK CORRELATION BETWEEN ,4A4,5H AND ,4A4,3H IS,F8.3)
      STOP
      END
```

ployed population, 1970; variable 2 is consumption of energy in kilograms of coal equivalent per inhabitants in 1971, and variable 3 is gross national produce in U.S. per inhabitant, 1970. (Source: J. P. Cole, *Geography of World Affairs,* Penguin Books, 1972.)

	Variable 1	Variable 2	Variable 3
Australia	8	5360	2820
Brazil	44	500	420
Ethiopia	85	30	80
India	68	190	110
Iran	46	900	380
Japan	21	3270	1920
Netherlands	6	5070	2430
Nigeria	67	60	120
Philippines	70	300	210
Spain	34	1610	1020
Turkey	69	520	310
U.K.	3	5510	2270
U.S.A.	4	11240	4760
Venezuela	26	2520	980
Zaïre	78	80	90

(38) Modify the ranking algorithm to arrange the ranks from low to high, i.e. the smallest value is given a rank of 1, the next smallest a rank of 2, and so on. Deal with equal ranks in the same way as before, though a slight modification will be necessary.

(39) A large number of statistical methods do not rely on assumptions about the distributional form of the variables to be analyzed (Spearman's rank correlation coefficient is one). These methods are termed *non-parametric*. Many non-parametric tests use ranking procedures. An example is the Wilcoxon test, which is the non-parametric equivalent of the Student's t test. Given two paired samples, x1 and x2, each of size n, the null hypothesis that the means $\overline{x1}$ and $\overline{x2}$ are equal is set up. Next, the differences X(I) between corresponding observed values of x1 and x2 are found. Differences of zero are not recorded. Rank the absolute values of these differences from low (1) to high (n) (do this by ranking ABS(X(I)) rather than X(I)). Treat tied values as explained in the modified ranking algorithm. Now add separately the ranks for the situations in which the x1 value is larger than the corresponding x2 value, and for those situations when the x2 value exceeds the corresponding x1 value. Call the smaller of the two sums T. Now compute

$$z = \frac{\frac{1}{4}n(n+1) - T}{\{n(n+1)(2n+1)/24\}^{\frac{1}{2}}}$$

where n is the number of observations in each sample. (Note: n must be at least 30 for this formula to be valid.) Look up a table of the standard normal distribution and find the computed z value. The probability (p) can then be read off. For example, if the computed z is 2.8, $p = 0.005$; if $z = 1.0$, $p = 0.317$. The probability, p, is the likelihood of finding a z value as high as the computed z value given that the null hypothesis is true. If p is small (less than 0.05 for a 95% significance level) then the null hypothesis is rejected.

Work out the value of z for the following set of data, which gives the percentage of people over 65 in each of 30 towns for (a) 1961 and (b) 1971. Now write a computer program to carry out the Wilcoxon test, and use the same data as input. The null hypothesis to be tested is that the mean percentage of over-65's is the same in the two populations from which this sample of 30 towns was drawn.

town:	1	2	3	4	5	6	7	8
1961:	10	15	12	14	6	18	9	3
1971:	9	16	10	15	6	19	8	4

town:	9	10	11	12	13	14	15	16
1961:	21	16	9	17	5	11	11	9
1971:	19	16	10	17	8	9	10	9

town:	17	18	19	20	21	22	23	24
1961:	8	12	13	21	16	19	6	8
1971:	9	10	12	22	17	18	7	8

town:	25	26	27	28	29	30
1961:	16	13	17	11	9	10
1971:	15	16	16	11	8	11

4.4 Simulation: Drainage Networks

Liao and Scheidegger (1968) describe a method of simulating drainage patterns on a digital computer. These authors point out that, in some instances, it is the number of connections between the stream confluences rather than the lengths of the stream segments that are of interest, for example in the calculation of bifurcation ratios. (The bifurcation ratio is the ratio of the number of stream segments of order n to the number of segments of order (n + 1).) In order to predict expected values of the number of stream segments of various orders and values of the bifurcation ratio, resort is made to simulation of the patterns that a stream network could possibly take.

A drainage network is a special kind of graph called an arborescence. (A graph is simply a set of straight lines joining a number of points; the lines are called edges and the junction points are called nodes—see Haggett and Chorley, 1969.) The number of arborescences with a given number of stream sources can be calculated. If it is assumed that every one of these possible arborescences is equally probable then expected values of the number of edges of a given order can be found by taking an average.

As an example, consider Figure 4.3 which shows an arborescent network with 7 sources.

Figure 4.3

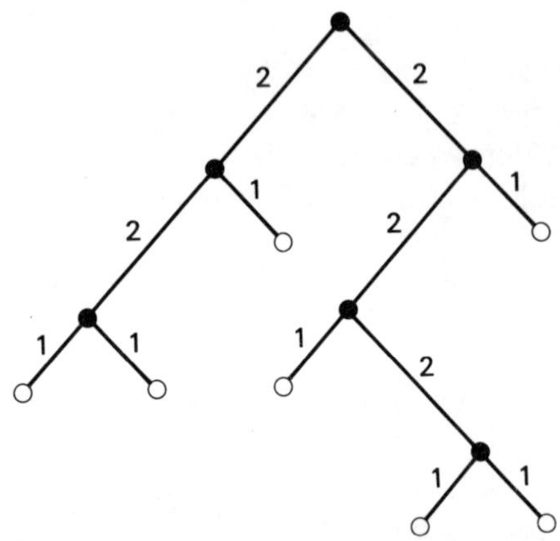

Figure 4.3 is a topological representation of the actual drainage pattern. Since we are interested only in the number of edges of a given order, the actual shape of the map pattern is irrelevant, as long as the pattern of connections is maintained. In Figure 4.3 a junction (node) is shown by the symbol ●, while a source is shown by the symbol ○.

It is impossible to get the computer to deal directly with a pattern. The computer works with numbers, so in order to simulate a pattern the pattern must first be converted into a set of numbers. This is relatively simple in the case of arborescences; the junctions (and the sink) are numbered with 1's and the sources with 0's. (Figure 4.4.) Now, starting at the sink, write down in order the 1's and 0's you pass when executing the following rule: work downwards, turning right whenever possible (when viewed from the sink) and writing down any 0's or 1's you pass. When a source

Figure 4.4

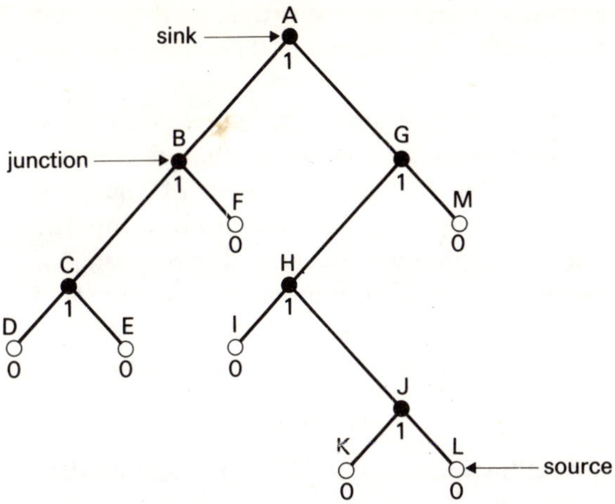

is reached go to the last preceding junction without writing down any 1's or 0's and turn left if you have not already done so. The network of Figure 4.4 can be converted to a number using the above procedure by following the following path:

A→B→C→D (→C)→E (→B)→F (→A)→G→ →H→I (→H)→J→K (→J)→L (→J→H→ →G)→M

giving the numerical sequence

1 1 1 0 0 0 1 1 0 1 0 0 0

The bracketed parts of the path indicate backward movement, when labels are not written down.

A second example is the network shown in Figure 4.5. Try to work out the numerical representation of this graph using the rule set out above. If you try several examples that

Figure 4.5

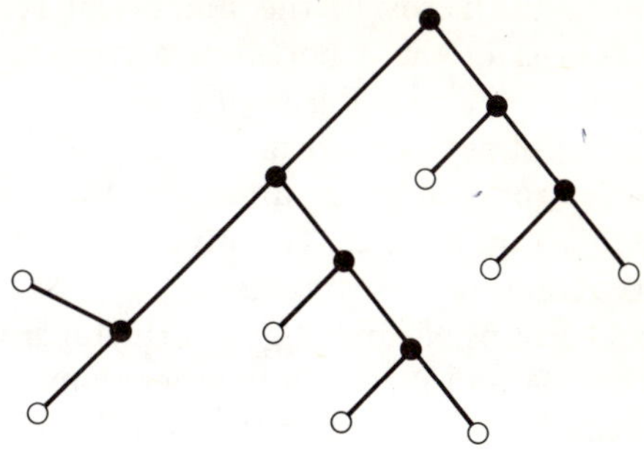

you make up yourself, you will find that:

(a) the number always starts with a 1 and ends with two 0's;
(b) there are always an odd number of digits, 2n – 1, where n is the number of sources;
(c) of these, n are zeros and n – 1 are 1s.

To work out the morphometric properties of the network, the number of segments of each order are required. This can be done visually if the number of sources is low (less than 6). For example, there are five possible networks with 4 sources. These are shown in Figure 4.6.

Taking network (d) in Figure 4.6, 4 first order segments can be counted together with one second order segment. (A segment is defined as a section between adjacent nodes. For example there are two second order segments in Figure 4.6(c), and only one second order segment in Figure 4.6(e).)

An algorithm is available to allow the derivation of the number of segments of each order, using the numerical representation of the graph. This algorithm is given by the

Figure 4.6

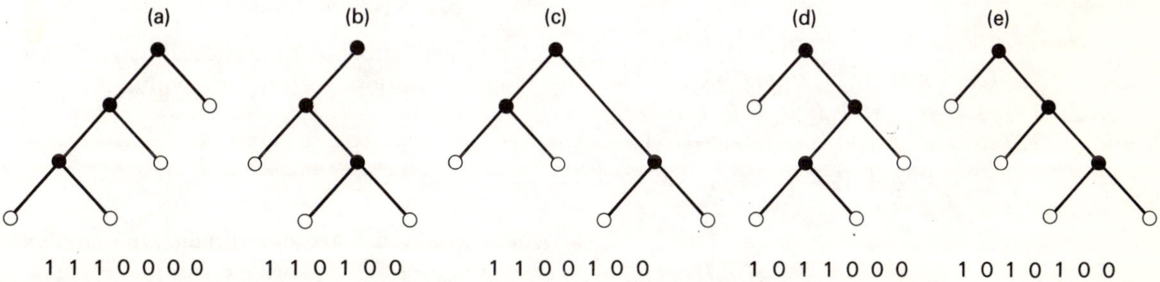

following four steps:

1. Count the zeros in the numerical representation of the arborescence; this is the number of first order segments.
2. Find groups of the form 100. Replace each such group of three digits by the single digit 2. Count the 2s. This gives the number of second order segments.
3. Find groups of three digits of the form 1AB where A and B are digits other than 1, for example 102 or 120 but not 112 or 101. Then:
   (a) If A = B replace the group of three digits by the single digit (A + 1) and count as one segment of order (A + 1).
   (b) If A is not equal to B, replace the group by the greater of A and B.
4. Keep executing Step 3 until only one digit remains. This gives the order of the drainage network.

EXAMPLES

Take the numerical equivalent of Figure 4.6(d):

1 0 1 1 0 0 0

Step 1: There are 4 zeros (n(1) = 4).
Step 2: 1 0 1 1 0 0 0
1 0 1 2 0 (n(2) = 1).
Step 3: 1 0 1 2 0
1 0 2
2.

Therefore, the number of first-order segments is 4, of second order segments is 1, and the order of the network is 2.

As a second example, take the numerical equivalent of Figure 4.5:

1 1 1 0 0 1 0 1 0 0 1 0 1 0 0

Step 1: There are 8 zeros (n(1) = 8).
Step 2: 1 1 1 0 0 1 0 1 0 0 1 0 1 0 0
1 1 2 1 0 2 1 0 2 (n(2) = 3).
Step 3: 1 1 2 1 0 2 1 0 2
1 1 2 2 2
1 3 2
3 (n(3) = 1).

For this network there are 8 first-order segments, three second-order segments and one third-order segment. The network has order 3.

The computer program for this exercise introduces one new concept, that of the subroutine. The fundamental idea of the subroutine is the same as that underlying the function, which was described in Section 3.5. A function is an independent section of a Fortran program that carries out a particular operation. For example, the statement

X = SQRT(Y)

causes the square root of the value stored in location Y to be placed in location X. The term Y in the statement is called an argument. Fortran compiler libraries contain many functions, some of which were listed in Section 3.5 (Table 3.14).

The actual value returned by the function (the square root of the contents of location Y in the last example) is held in a storage location which has the same name as the function (SQRT in the example). Subroutines, on the other hand, do not have a storage location corresponding to the subroutine name. The results of calculations performed in the subroutine are returned as arguments to the call instruction. For example, a subroutine can be called INDIC even though the values it returns as arguments are of type floating point.

Subroutines are just like straightforward Fortran programs, except:

(i) they begin with a SUBROUTINE statement which precedes the DIMENSION statement (if any), and
(ii) they end with a RETURN statement instead of a STOP statement.

The use of subroutines is best illustrated by example. Suppose we wish to evaluate

$$y = \sum_{i=1}^{20} x_i + \sum_{i=1}^{20} a_i + \sum_{i=1}^{20} z_i$$

where x, a and z are one-dimensional arrays (called vectors in algebraic terminology). The summation (Σ) operation is carried out a total of three times—once for x, once for y and once for z. A program to do this is shown in Table 4.6. Notice that the summation operation

Table 4.6

```
   DIMENSION X(20),A(20),Z(20)
   READ(5,1) (X(I),I=1,20)
   READ(5,1) (A(I),I=1,20)
   READ(5,1) (Z(I),I=1,20)
 1 FORMAT(20F4.0)
   Y1=0.0
   DO 2 I=1,20
 2 Y1=Y1+X(I)
   Y2=0.0
   DO 3 I=1,20
 3 Y2=Y2+A(I)
   Y3=0.0
   DO 4 I=1,20
 4 Y3=Y3+Z(I)
   Y=Y1+Y2+Y3
   WRITE(6,5) Y
 5 FORMAT(10HORESULT = ,F10.4)
   STOP
   END
```

is programmed three times—in DO loops 2, 3 and 4. A single subroutine could be written to carry out this operation and 'called' each time summation of a vector was needed. (Table 4.7.) The subroutine is called three

Table 4.7

```
   DIMENSION X(20),A(20),Z(20)
   READ(5,1) (X(I),I=1,20)
   READ(5,1) (A(I),I=1,20)
   READ(5,1) (Z(I),I=1,20)
 1 FORMAT(20F4.0)
   CALL SUM (X,Y1)
   CALL SUM (A,Y2)
   CALL SUM (Z,Y3)
   Y=Y1+Y2+Y3
   WRITE(6,5) Y
 5 FORMAT(10HORESULT = ,F10.4)
   STOP
   END
```

```
   SUBROUTINE SUM (A,S)
   DIMENSION A(20)
   S=0.0
   DO 1 I=1,20
 1 S=S+A(I)
   RETURN
   END
```

times by a CALL statement. The CALL statement is of the form

CALL subroutine name (argument list)

while the SUBROUTINE statement that begins the subroutine is of the form

SUBROUTINE subroutine name (argument↔
↔list)

There must be the same number of arguments in the argument lists of the CALL and SUBROUTINE statements, and corresponding arguments must be of the same type (i.e. floating point or integer). The actual names of the argument need not agree, as the above example shows (Table 4.7). At the time the subroutine is called the first argument in the SUBROUTINE statement is linked to the first argument in the CALL statement, and so on. A subroutine can therefore be used in any program since the actual argument names are not important, except insofar as they indicate the type (floating point or integer) of the variable or array.

If the dimensions of the arrays used in subroutines had to match exactly the dimensions of the corresponding arrays in the calling program then the usefulness of the subroutine would be limited. It is possible to set up the dimensions of the arrays used within a subroutine when the subroutine is actually called. For example:

```
DIMENSION X(75)
. . . . . . . . . . . . . . .
. . . . . . . . . . . . . . .
CALL SUBR(X,75)
. . . . . . . . . . . . . . .
STOP
END
```

(continued)

```
SUBROUTINE SUBR(X,N)
DIMENSION X(N)
.................
.................
RETURN
END
```

The value 75 in the CALL statement is associated with the location N in the SUBROUTINE statement. Thereafter, within the subroutine, the contents of location N will be 75, even in the dimension statement. Thus, the same subroutine could be used in another program as follows:

```
DIMENSION Y1MP(100)
.....................
.....................
CALL SUBR(Y1MP,100)
.....................
STOP
END
SUBROUTINE SUBR(X,N)
DIMENSION X(N)
...............
...............
RETURN
END
```

In this case, the contents of location N in the subroutine will be 100, not 75 as before. The only proviso is that the dimension of an array named as an argument in a SUBROUTINE statement cannot be greater than the dimension of the corresponding array named in the CALL statement.

It is always good practice to put the value of any constants in store locations (such as N1, N2) and use the names of those locations as arguments in CALL statements, rather than use actual numerical values as arguments.

EXERCISE

(40) Write a subroutine to add together all the elements of a one-dimensional array and return the sum to the calling program. Write two calling programs, one to read in an array of 25 floating point numbers and find the sum of the 25 elements by calling the subroutine, and the second to read in an array of 10 floating point numbers and treat it in a similar way.

Example Program 14 uses one subroutine—called REORG—to move all the digits to the right of a '1' two places left when a '100' or '1AB' has been found (step 2 and step 3 of the algorithm). It is called every time a group of digits of the form '100' or '1AB' is found. Go through the example program and see if you can follow its workings. Refer to the algorithm above, and note that the integer Fortran library function MAX0(I1,12. . .,IN) returns the value of the largest integer in the list (I1,12,. . . . .IN).

Table 4.8

```
C SET UP ARRAY DIMENSIONS. *NUMBER* IS THE BINARY REPRESENTATION OF THE
C NETWORK AND *NSEGS* IS THE NUMBER OF STREAMS (LINKS) OF EACH ORDER.
      DIMENSION NUMBER(100),NSEGS(10)
C READ IN THE NUMBER OF EDGES (= SOURCES + JUNCTIONS)  AND STORE IN *N*
C ALSO READ THE BINARY REPRESENTATION INTO *NUMBER*.
      READ(5,1) N
      N2=N-2
    1 FORMAT(I4)
      READ(5,2) (NUMBER(I),I=1,N)
    2 FORMAT(80I1)
      WRITE(6,20) N, (NUMBER(I),I=1,N)
   20 FORMAT(22HONUMBER OF SEGMENTS = ,I4//17HOBINARY NUMBER = ,80I1)
   21 FORMAT(1H ,80I1)
```

Table 4.8 (continued)

```
C SET *NSEGS* TO ZERO.
      DO 22 I=1,10
   22 NSEGS(I)=0
C STEP 1 COUNT THE ZEROS, PUTTING THE TOTAL INTO NSEGS(1).
      DO 3 I=1,N
    3 IF(NUMBER(I).EQ.0) NSEGS(1)=NSEGS(1)+1
C STEP 2: NOW LOOK FOR GROUPS OF THREE OF THE FORM '100'.
      I=0
    4 I=I+1
      IF(I.GT.N2) GOTO 5
      IF(NUMBER(I).NE.1) GOTO 4
      I1=I+1
      I2=I+2
      IF((NUMBER(I1).NE.0).OR.(NUMBER(I2).NE.0)) GOTO 4
C FOUND ONE! INCREASE NSEGS(2) BY 1, THEN REPLACE THE '100' BY '2'.
C NEXT, MOVE ALL DIGITS ABOVE THE '2' TWO PLACES LEFT,TO FILL THE SPACE
C PREVIOUSLY OCCUPIED BY THE '00' PART OF THE '100'. (THE '2' REPLACES
C THE '1').
      NSEGS(2)=NSEGS(2)+1
      NUMBER(I)=2
      CALL REORG(I,N,N2,NUMBER)
      WRITE(6,21) (NUMBER(J),J=1,N)
      GOTO 4
    5 IF( N.EQ.1) GOTO 11
C STEP 3
C NOW LOOK FOR GROUPS OF THE FORM '1AB' (NEITHER A NOR B EQUAL TO 1).
   14 I=0
    7 I=I+1
      IF(I.GT.N2) GOTO 11
      IF(NUMBER(I).NE.1) GOTO 7
      I1=I+1
      I2=I+2
      IF((NUMBER(I1).EQ.1).OR.(NUMBER(I2).EQ.1)) GOTO 7
C FOUND ONE! NOW TEST FOR A=B AND ACT ACCORDINGLY -- STEP 3A OR STEP 3B.
      IF(NUMBER(I1).EQ.NUMBER(I2)) GOTO 8
C MAXO IS A FUNCTION WHICH GIVES THE INTEGER VALUE OF THE LARGEST
C INTEGER ARGUMENT ---SEE TEXT.
      L=MAXO(NUMBER(I1),NUMBER(I2))
      NUMBER(I)=L
      GOTO 9
    8 L=NUMBER(I1)+1
      NSEGS(L)=NSEGS(L)+1
      NUMBER(I)=L
    9 CALL REORG(I,N,N2,NUMBER)
      WRITE(6,21) (NUMBER(J),J=1,N)
      GOTO 7
   11 IF(N.EQ.1) GOTO 13
```

Table 4.8 (continued)

```
      GOTO 14
C FINISHED.  NOW OUTPUT RESULTS.
   13 NORD=NUMBER(1)
      WRITE(6,12) NORD, (I, NSEGS(I),I=1,NORD)
   12 FORMAT(25HOORDER OF THIS NETWORK = ,I4/19HOSTREAM NUMBERS ARE/
     1 10(1H ,I3,11HTH ORDER = ,I4/))
      STOP
      END


      SUBROUTINE REORG(I,N,M,NUMBER)
C THIS SUBROUTINE MOVES ALL DIGITS TO THE RIGHT OF THE ITH TWO
C PLACES LEFT
C N IS THE NUMBER OF DIGITS IN THE NUMBER
C M IS EQUAL TO N-2
C NUMBER IS AN INTEGER ARRAY HOLDING THE NUMERICAL REPRESENTATION OF
C THE NETWORK
      DIMENSION NUMBER(100)
      J=I+1
      IF(J.GT.M) GOTO 2
      DO 1 K=J,M
      L=K+2
    1 NUMBER(K)=NUMBER(L)
    2 N=N-2
      M=M-2
C N AND M ARE REDUCED BY 2 BECAUSE TWO DIGITS HAVE BEEN REMOVED
      RETURN
      END
```

EXERCISE

(41) Punch out and run Example Program 14 using as data the numerical representation of the arborescences shown in Figure 4.6 (a) to (e).

A simulation program is not of much use if the range of possible values has to be read in as data. It would be exceedingly tedious to work out the numerical representation of all arborescences with more than 7 or 8 sources. One method of finding all the possible arborescences with n sources would be to generate all possible combinations of 1s and 0s subject to the restrictions (a) to (d) set out below. Such an approach would be very expensive if n became large, for example if $n = 17$ then the total number of possible arborescences is 35,357,762. If the computer took 1 second to determine the sequence of 1s and 0s and to deal with it according to the algorithm outlined above, it would take around 58 weeks of computing time to analyze all the different arborescent networks. This is obviously infeasible. Statistical methods of randomly sampling a population have been developed to deal with situations in which the total population is either unavailable or too large to be treated as a whole. In order to use random sampling methods in the simulation example, it is necessary to assume that each of the possible networks is equally likely. Also, a method of sampling must be worked out. The remarks on page 65 hold the key to a method of getting the computer to work out a random sample. The numerical representation (a) always starts with a 1 and ends with two 00s; (b) the number of digits is ntot = (2n − 1) where n is the number of sources; (c) of these, n are zeros and (n − 1) are ones, and (d) counting from the left, the cumulative number of zeros must not exceed the cumulative number of ones. Thus, the sequence

100011100

is illegal, because at digit 3 there are (counting from the left) 2 zeros and only 1 one. Similarly, digit 4 is illegal, because there are (cumulatively) 3 zeros and 1 one. This restriction is necessary to prevent more than two streams joining at any one junction, as the algorithm cannot deal with such cases.

The ratio of the number of 1s to the total number of unallocated digits is (n − 1 − nl)/nleft where nl is the number of 1s already allocated, (n − 1) is the total number of ones, and nleft is the total number of digits left to fill. Digits are allocated from left to right—i.e. digit 1 is shown first, digit 2 next, and so on. However, the first digit must always be 1 and the last two digits must be 00, so these digits can be allocated immediately. Thus, if there are n = 4 sources, there are ntot = (2n − 1) = 7 digits in the numerical representation. Digit 1 must be a 1, and digits 6 and 7 must both be 0. The numerical representation is thus of the form 1. . . . 00, where the symbol '.' stands for an unallocated digit. The number of 1s is (n − 1) = 3; nl is currently equal to 1, so the ratio (n − 1 − nl)/nleft = (4 − 1 − 1)/4 = 0.5 for digit 2. If digit 2 was chosen to be a 1, then nl would become 2, nleft would become 3 and the ratio would be (4 − 1 − 2)/3 = 0.3333̇. As another example, take the incomplete numerical representation 1. . . . . 00. The values of nl and nleft are 1 and 5 respectively, so (n − 1 − nl)/nleft = 2/5 = 0.4. This means that the probability of the second digit being a 1 is 2/5; the probability that it is 0 is therefore 3/5. A random selection of a card from a set of five cards, two of which indicate a 1 and three a 0, would provide a means of choosing. Imagine that the choice produces a 1, so the number becomes 11 . . . . 00. The ratio of the number of 1s remaining to the number of unallocated digits is now 1/4 so we must choose a card from a set of four, only one of which indicates a 1. The procedure is carried out for all unallocated digits until the numerical representation is complete.

The method of selecting cards can be simulated by the use of a pseudo-random number generator. The numbers generated appear to be random, but they are not truly random because they are calculated according to a set formula. All computing centres or computer manufacturers provide a Fortran function to generate pseudo-random numbers, for example, PDP-11 computers have a function called RAN; CDC 6600/7600 computers have a similar function called RANF. You should check with your computing centre to find the name of the corresponding function for your computer. The pseudo-random numbers generated by these functions are distributed uniformly between 0 and 1. That is to say, the probability of selecting any number r in the range $0 \leqslant r \leqslant 1$ is equally likely. The number 0.5000000 is as likely to occur as 0.1234567 if the machine works to seven significant decimal digits. It can be deduced from this that the probability of a random number less than 0.5 being generated is 0.5, the probability of a random number less than 0.6161071 being generated is 0.6161071, and so on. This fact can be used to simulate the drawing of cards from a pack. To simulate a situation in which the pack contains five cards, two of which indicate 1s, the other three indicating 0s, firstly work out the probability of a card indicating a 1 being selected. The probability is 0.4. Devise a rule as follows: if the pseudo-random number generator produces a random number less than 0.4, then a 1 has been drawn. If the random number is 0.4 or more then a 0 has been drawn. The probability will, of course, alter when the 1 or 0 has been added to the numerical representation.

Using these facts, and the restrictions (a)

to (d) on the form of the numerical representation of the arborescence, a subroutine (GENER) has been written (Table 4.9). It will fill in the unallocated digits of a numerical representa-

Table 4.9

```
C SET UP ARRAY DIMENSIONS. *NUMBER* IS THE BINARY REPRESENTATION OF THE
C NETWORK AND *NSEGS* IS THE NUMBER OF STREAMS (LINKS) OF EACH ORDER.
      DIMENSION NUMBER(100),NSEGS(10)
C FIRST DATA CARD CONTAINS NUMBER OF SOURCES (INTEGER - COLS 1-3)
C AND NUMBER OF NETWORKS TO BE SIMULATED (INTEGER - COLS 4-6).
      READ(5,1) NSRC, NTIMES
    1 FORMAT(2I3)
C II1 AND II2 ARE USED IN THE PDP-11 PSEUDO-RANDOM NUMBER GENERATOR
C YOU WILL PROBABLY NEED TO INITIALISE SIMILAR VARIABLES
C FOR THE COMPUTER SYSTEM YOU ARE USING.
      II1=0
      II2=0
      DO 1000 KT=1,NTIMES
C N IS THE TOTAL NUMBER OF 1'S AND 0'S.
C N2 IS THE IDENTIFIER OF THE HIGHEST UNALLOCATED DIGIT (COUNTING
C FROM THE LEFT).
C NONES WILL COUNT THE NUMBER OF ONES.
C NZEROS WILL COUNT THE NUMBER OF ZEROS.
      N=NSRC+NSRC-1
      N2=N-2
      NONES=1
      NZEROS=0
C NOW SET THE FIRST DIGIT TO 1 AND THE LAST TWO DIGITS TO 0.
      NUMBER(1)=1
      NUMBER(N-1)=0
      NUMBER(N)=0
C NLEFT WILL COUNT THE NUMBER OF UNALLOCATED DIGITS.  THREE DIGITS
C HAVE BEEN ALLOCATED SO FAR.
      NLEFT=N-3
C NN IS THE POSITION OF THE LAST (RIGHTMOST) UNALLOCATED DIGIT.
C THE FOLLOWING DO-LOOP ALLOCATES DIGITS 2 TO NN USING A PSEUDO-
C RANDOM NUMBER GENERATOR AS EXPLAINED IN THE TEXT.
      NN=NLEFT+1
      DO 33 I=2,NN
C RATIO IS THE RATIO OF THE NUMBER OF ONES TO BE ALLOCATED TO THE
C NUMBER OF UNALLOCATED DIGITS REMAINING.
      RATIO=FLOAT(NSRC-NONES-1)/FLOAT(NLEFT)
C GENER IS THE SUBROUTINE TO GENERATE THE 1'S AND 0'S.
   33 CALL GENER(NUMBER,N,NLEFT,NSRC,II1,II2,NONES,NZEROS,
     1 RATIO,I)
      WRITE(6,20) NSRC, (NUMBER(I),I=1,N)
   20 FORMAT(22HONUMBER OF SEGMENTS = ,I4//17HOBINARY NUMBER = ,80I1)
   21 FORMAT(1H ,80I1)
C SET *NSEGS* TO ZERO.
```

• • • • • • • •

Table 4.9 (continued)

```
      SUBROUTINE GENER(NUMBER,NDIM,NLEFT,N,I1,I2,NONES,NZEROS,RATIO,I)
C ARGUMENTS ARE:
C NUMBER    ARRAY HOLDING NUMERICAL REPRESENTATION OF ARBORESCENCE
C NDIM      TOTAL NUMBER OF DIGITS
C NLEFT     NUMBER OF UNALLOCATED DIGITS
C N         NUMBER OF SOURCES
C I1,I2     NEEDED BY PDP-11 RANDOM NUMBER GENERATOR.
C NONES     NUMBER OF 1'S ALLOCATED SO FAR.
C NZEROS    NUMBER OF 0'S ALLOCATED SO FAR
C RATIO     RATIO AS EXPLAINED IN TEXT
C I         NUMERICAL IDENTIFIER OF DIGIT BEING CONSIDERED
      DIMENSION NUMBER(NDIM)
C GENERATE RANDOM NUMBER
      X=RAN(I1,I2)
C IS RANDOM NUMBER (X) GREATER THAN RATIO?
      IF(X.GT.RATIO) GOTO 1
C IF IT IS, PUT A 1 INTO DIGIT I,AND INCREASE NONES BY 1
    5 NUMBER(I)=1
      NONES=NONES+1
      GOTO 2
C
C IF RANDOM NUMBER IS LESS THAN OR EQUAL TO RATIO, PUT A 0 IN
C THE ITH POSITION, UNLESS RULE (D) WOULD BE BROKEN - SEE TEXT.
    1 IF(NZEROS.EQ.NONES) GOTO 5
C IF NZEROS IS CURRENTLY EQUAL TO NONES THEN THE ADDITION OF ANOTHER
C ZERO WILL BREAK RULE (D) SO FORCE THE NEXT DIGIT TO BE A 1.
C OTHERWISE IT IS A 0, SO INCREASE NZEROS BY 1 AND
C REDUCE NLEFT BY 1.
      NUMBER(I)=0
      NZEROS=NZEROS+1
    2 NLEFT=NLEFT-1
C RETURN TO CALLING PROGRAM
      RETURN
      END
```

tion according to the rules already set out. The pseudo-random number generator employed is function RAN, the PDP-11 function. You will have to substitute your own version of RAN. Comments in the listing should help to explain the logic underlying the subroutine. An extract from a possible calling program is also shown in Table 4.9.

EXERCISES

(42) Attach subroutine GENER to Example Program 14, supplying the necessary CALL statement. (Refer to Table 4.9.) Use the composite program to draw samples of size 40 from the populations of arborescent networks with (a) 6, (b) 10 and (c) 20 sources. Work out the bifurcation ratios (or add Fortran instructions to Example Program 14 to do this for you).

(43) The number of possible arborescences with n sources is $(1/k).(k!/n!(k-n)!)$ where $k = 2n - 1$ and $k! = k \times (k-1) \times (k-2) \times \ldots \times 1$. (n! and (k − n)! are defined in a similar

way.) Write a Fortran program to evaluate this expression for n = 3 to n = 20 sources. Remember that the results will be very large integers. Most computers cannot store very large integers (for example, the limit for integers is 32,768 for PDP-11 computers, 2,147,483,647 for IBM System/360 computers and 8,388,607 for ICL 1900 Series computers). You may find, therefore, that it is necessary to do the calculations using floating point arithmetic, as the range of numbers representable in floating point form is considerably greater than in integer form. An IBM System/360 computer can hold floating point numbers in the range 10^{-78} to 10^{75} approximately, while an ICL 1900 Series computer can hold floating point numbers in the range -5.6×10^{76} to 5.6×10^{76}.

(44) The growth and interaction of populations—human and animal—has been the subject of several simulation models, some of which are summarized by Leinbach (1974, pages 54–66). The model considered here is the well-known prey—predator model. The predator in this exercise is the lynx (y) and the prey are hares (x), though they could be competing nations or alternative forms of transport. Assume that the lynxes and hares live in isolation, the lynx feeding only on hares. In the absence of lynxes, the hare population grows directly in proportion to its size, that is:

$$dx/dt = Ax$$

where A is an empirically-determined constant. The notation dx/dt means: the change in x (number of hares) per unit change in time. If there were no hares, the lynx population would decline over time as the lynxes died from starvation. Thus:

$$dy/dt = -Py$$

Again, P is an empirically-determined constant.

The sizes of the hare and lynx populations are related in that the number of hares eaten by each lynx is directly proportional to the number of hares. Since there are x hares and y lynxes, the number of hares eaten per unit of time is proportional to xy. If this interaction is taken into account, then the hare population will decline in direct proportion to xy and the lynx population will increase in direct proportion to xy. The basic prey—predator model is thus:

$$dx/dt = Ax - Bxy$$
$$dy/dt = Cxy - Py$$

Leinbach gives the following values for the empirically-determined constants: A = 0.93, B = 0.04, C = 0.004 and P = 0.22. Assuming that at the start of the period x = 100 (thousand) and y = 20 (thousand), write a Fortran program to simulate the interaction between the lynx and hare populations over a 50 year period. Evaluate the model at monthly intervals, but print the totals for x and y annually.

As a supplementary exercise, (i) modify the program so that an upper limit can be placed on the lynx population (y), i.e. at the end of each month y will be set to the minimum of (calculated y, specified upper limit); (ii) add a section of program to cause a graph of x and y against time to be drawn. You will need to use A1 format to do this, as follows: set the locations BLANK, FLYNX and HARE to the characters Δ, Y and X by means of a DATA statement which is described on page 82. Next, set all 100 elements of an array SLINE to BLANK. When, at the end of each month, the values of x and y are computed, find K1 = INT(x/150.0*99.0) + 1 and K2 = INT(y/150.0*99.0) + 1. This gives the relative positions of x and y along a scale from 0 to 100, assuming the maximum value either x or y can take is 150 (thousand). Now set SLINE(K1) to HARE and SLINE(K2) to FLYNX. Print out the contents of SLINE using 100A1 format, then set both SLINE(K1) and SLINE (K2) to BLANK. (You will need to check that neither K1 nor K2 exceeds 100; if so, set K1 (or K2) to 100.)

Run the program several times with different upper limits to the lynx population, and write a short description of your findings.

4.5 Linear Programming—The Simplex Method

One of the more beneficial of the results of the Second World War was the development of electronic computers, which was described in Chapter 2. Another outcome of the war was the invention of new methods of management, for the handling of large numbers of troops and equipment was a problem in logistics not previously faced even by the largest commercial companies. One such method of management is called Operations Research, which includes a variety of techniques. One of these techniques is Linear Programming. This is not programming in the sense of computer programming, but it can be thought of as scheduling or logical organization. Linear programming is a method of defining a complex problem and identifying the optimum strategy for its resolution. For example, a simple situation might involve the following facts: a farmer has a £200 weekly budget, and has a labour force of eight men, each working a 40-hour week, giving a total availability of 320 man-hours per week. The farmer can grow either wheat or sugar beet, or a proportion of each of these crops. The wheat requires an input of 8 man-hours per crop unit per week, and an expenditure of £4 per week on fertilizer, tractor fuel, and so on. The sugar beet is less demanding in terms of labour, needing only 5 man-hours per crop unit per week, but needs more fertilizer (£5 per week). The farmer can expect to make a profit of £1.75 per unit of wheat but only £1.50 per unit of sugar beet. How much of each crop should he grow?

The problem can be resolved graphically as follows (Figure 4.7). Draw a pair of perpendicular axes; label the vertical axis 'wheat' and the horizontal axis 'beet'. Now, take the information relating to labour requirement. The production of beet needs 5 man-hours per crop unit per week, whereas wheat needs 8 man-hours per unit per week, and a total of 320 man-hours per week is available. Thus, given the available labour-force the farmer could grow, at one extreme, 320 ÷ 8 = 40 units of wheat or, at the other extreme,

Figure 4.7

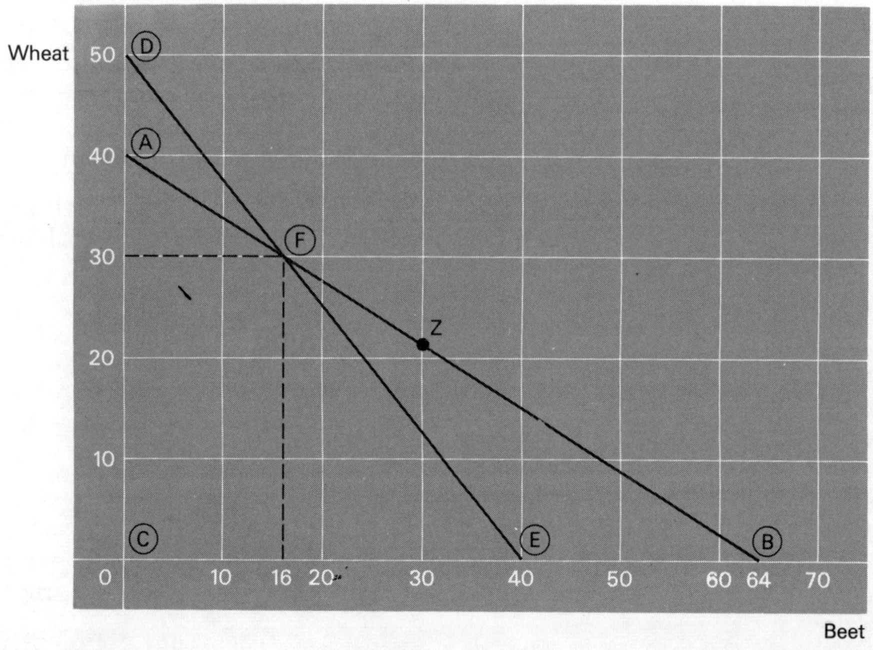

$320 \div 5 = 64$ units of beet, assuming that he has enough land and that all the men are to be fully employed. Locate the point 64 on the horizontal (beet) axis and the point 40 on the vertical (wheat) axis, and join the two points to give the line AB. The fact that A and B are joined by a straight line is the reason why the technique is called linear programming. The line joins all points which represent feasible allocations of labour, for example the point Z gives 30 units of beet and 20.5 units of wheat which uses $(30 \times 5) + (20.5 \times 8) = 314$ man-hours, which is close enough to the available maximum (320) considering that rough graphical methods were used. In addition, any point lying within the triangular area enclosed by the two axes and the line AB is feasible, that is, possible given the limitation of labour available, but total production would not be maximized. It is possible, for example, to select the point (0,0), since this lies within or on the boundary of the triangle ABC, but this implies that no crops would be grown at all, no labour used, and no profit made. The optimum of labour is represented by a point somewhere within or on the boundary of the triangle ABC in Figure 4.7, but before the position of the optimum point can be identified the information regarding cost must be used. Again, the farmer could spend his £200 exclusively on wheat (giving $200 \div 4 = 50$ units of wheat) or exclusively on beet (giving $200 \div 5 = 40$ units of beet). Using the same logic as previously, all feasible points must lie within or on the boundary of $\triangle$DEC, as far as monetary cost is concerned. Looking at $\triangle$DEC and $\triangle$ABC, their common area is the quadrilateral AFEC, so the (joint) optimum point must lie somewhere within or on the boundary of this quadrilateral.

One way to find the optimum point would be to evaluate the amount of profit made at each of a set of points within the quadrilateral AFEC. For example, at point A forty units of wheat and no beet would be grown, giving a profit of £$(40 \times 1.75) =$ £70. Then the point of maximum profit could be identified and the proportions of wheat and beet read off the relevant axes. This approach could work in a relatively simple case, such as in this example, but where more than two crops were involved the strategy would be much more complex. In fact, all graphical solutions will work only in simple cases. Without attempting to show why, it will be stated as a fact that the point of maximum profit (optimum point) in the present case is the point F, giving 30 units of wheat and 16 of beet, and a total profit of £76.50.

A mathematical approach to the same problem is possible. The advantage of the mathematical approach is that it can be generalized to include many variables (crops) and constraints (costs, labour requirements) but its disadvantage is the complexity of the calculations. The computing algorithm is an iterative one, involving repetition of a small set of instructions, so that programming the method is relatively straightforward. The steps are:

(i) Write down the information relating the variables (e.g. the crops) and the constraints (e.g. the cash and man-hour limitations). For the wheat (w) and beet (b) problem this gives:

$$8w + 5b \leqslant 320$$
$$4w + 5b \leqslant 200$$

where the sign $\leqslant$ means: 'is less than or equal to', and w and b stand for wheat and beet. These two relationships are termed inequalities because the left and right hand sides are connected by a $\leqslant$ sign.

(ii) Write down the equation giving the profit function, P, in terms of the variables:

$$P = 1.75w + 1.5b$$

(iii) Lay out the following table:

w	b	x1	x2	x3
8	5			320
4	5			200
−1.75	−1.5	0	0	0

Row 1 of the table holds the coefficients of the first inequality at step (i), with the restriction (320 man-hours) in the column headed x3. Columns x1 and x2 are empty at the moment. Row 2 holds the coefficients of the second inequality, and Row 3 the coefficients of the profit function (step (ii)) but with the signs reversed. The entries x1 to x3 in Row 3 are zero. The empty positions in this table are filled in as follows: since (8w + 5b) has to be less than or equal to 320 man-hours, then let the difference between (8w + 5b) and 320 be represented by the quantity x1. Now it is possible to write $8w + 5b + x1 = 320$. This is an equality (that is, it is a mathematical statement involving an $=$ sign rather than a $\leqslant$ or $\geqslant$ sign). The actual value of x1 is of no concern—it is simply the difference between (8w and 5b) and 320. It can be considered as a variable which 'takes up the slack' (i.e. unused) portion of a particular allocation, hence x1 is called a slack variable. A second slack variable x2 is introduced into row 2 to give $4w + 5b + x2 = 200$. The two equations are:

$$8w + 5b + 1x1 + 0x2 = 320$$
$$4w + 5b + 0x1 + 1x2 = 200$$

The coefficients can be written in matrix form as in Table 2.

8	5	1	0	320
4	5	0	1	200
−1.75	−1.5	0	0	0

The (negative) coefficients of the profit function (step (ii)) make up row 3. This table is simply Table 1 with the empty positions filled by the slack variables.

(iv) Choose the largest negative value in row 3 of Table 2, and note its column number. (In this case it is −1.75, lying in column 1.)

(v) Find the result of dividing the elements of the last column by those of the column found at step (iv)—in this case, column 1. Choose the smallest positive value, and note its row number. The result of the divisions in the example is $320 \div 8 = 40$, $200 \div 4 = 50$ and $0/-1.75 = 0$. The smallest positive value is 40, lying on row 1.

(vi) The number lying at the intersection of the column found at step (iv) and the row found at step (v) is called the pivot element. In the example the pivot element is element (1,1) and has the value 8. The row containing the pivot element is the pivot row, and the column in which the pivot element lies is called the pivot column.

(vii) Divide all the elements on the pivot row by the pivot element to give the new pivot row with elements $p_1, p_2, p_3, \ldots$. If the index (number) of the pivot column is represented by K, then the remaining elements in Table 2 are transformed by the use of equation (1):

$$r_j^* = r_j - r_k \cdot p_j \qquad (1)$$

where r_j is the jth element of the (non-pivot) row that is being transformed, r_j^* is the corresponding transformed value, r_k is the (untransformed) k^{th} element of the row being transformed—i.e. it is the element of the current row lying in the pivot column—and p_j is the value of the jth element of the new pivot row. The result of this operation should be to alter all the elements of the table, making the pivot element equal to one and all other elements in the pivot column equal to zero. The pivot row in Table 2 is row 1, so divide each element of row 1 by 8 (the pivot element) to obtain $p_1 = 1$, $p_2 = 0.625$, $p_3 = 0.125$, $p_4 = 0$ and $p_5 = 40$. Now transform all the other elements of Table 2 by the use of formula (1). Row 2, for example, is transformed by noting that $K = 1$ and $r_k = 4$, and then finding

$$r_1^* = 4 - 4 \times 1 = 0$$
$$r_2^* = 5 - 4 \times 0.625 = 2.5$$
$$r_3^* = 0 - 4 \times 0.125 = 0.5$$
$$r_4^* = 1 - 4 \times 0 = 1$$
$$r_5^* = 200 - 4 \times 40 = 40$$

EXERCISE

(45) Calculate the r^* values for row 3 of the transformed table.

Table 3 can now be written out by assembling

the elements of the new pivot row (the p_i) and the transformed elements of rows 2 and 3 (the r^*).

1	0.625	0.125	0	40
0	2.500	−0.500	1	40
0	−0.406	0.220	0	70

Table 3 is interpreted as follows. Since there are two variables (crops) in this problem, look at columns 1 and 2 only. If one of these columns starts with a 0 or a 1, locate the number of the row on which the 1 lies and look up the value on the same row and in the last column. This gives the number of units of that variable produced. If a column does not begin with a 0 or 1, then no units of the variable concerned are produced at this stage. Thus, at the end of the first cycle, 40 units of crop 1 (wheat) are produced. The profit on wheat is £1.75 per unit, and the total profit is thus $£(1.75 \times 40) = £70$. This is the value in the last row and last column. No units of beet are grown at this stage.

If none of the elements of the last row is negative this is the optimum profit. If there are any negative elements, return to step (iv) and repeat steps (iv), (v), (vi) and (vii) until all the elements in the last row are positive or zero. The positions of the 1s (if any) in the columns corresponding to real variables (crops) give the number of units produced.

Since there is a negative element in the last row of Table 3, a second cycle (steps (iv)–(vii)) is begun with the location of the largest negative element in the last row (−0.406 in column 2), and the division of the elements of the final column by the elements of column 2. The smallest positive value in these dividends gives the row number of the pivot element. The dividends are $40 \div 0.625 = 64$, $40 \div 2.5 = 16$ and $70 \div -0.406$, which we needn't evaluate, as it is nonpositive. The pivot element is thus on column 2 and row 2, i.e. it is 2.5.

Table 3 is now transformed to give Table 4 in the same way that the operations of step (vii) produced Table 3 from Table 2 by dividing the elements of the pivot row at this iteration (row 2) by 2.5, the value of the pivot element, and then using Formula 1 of step (vii) to transform the rest of Table 3 into Table 4.

1	0	0.250	−0.250	30
0	1	−0.200	0.400	16
0	0	0.727	0.615	76.50

The table has no negative elements in its final row, and therefore represents the optimum solution. The values (in crop units) of wheat and beet are $w = 30$ and $b = 16$, and the maximum profit is £76.50—the same as that found graphically. The values $w = 30$ and $b = 16$ can be substituted into the original inequalities:

$$8w + 5b \leqslant 320 \text{ (man hours)}$$
$$4w + 5b \leqslant 200 \text{ (pounds Sterling)}$$

The result is that $8w + 5b = 320$, so all the workers are kept fully employed, and $4w + 5b = 200$, so the farmer spends all his capital in achieving the profit of £76.50.

Not all linear programming problems have solutions in which the constraints are satisfied exactly. Nor is it necessary—as in the example just given—to have exactly as many variables (crops) as constraints. It is possible to add further constraints. The next example will illustrate this situation.

A fictional country, Primonia, has three basic products—sugar (x), rice (y) and groundnuts (z). The total available capital outlay per annum in Primonia is \$45 million; the capital investment required to produce one unit of the three crops is \$3m, \$6m and \$4m respectively. The first inequality is then

$$3x + 6y + 4z \leqslant 48$$

The population of Primonia is such that a total of 136,000 man-hours will be available, under normal circumstances, in a given year. The labour requirements of each crop (in thousands of man hours) are 11, 6 and 12, hence the second inequality can be written

$$11x + 6y + 12z \leqslant 136$$

The third restriction is that of transporting

capacity. There are 70 transportation units available to move crops, seed and labour, and the requirements can be expressed by

$$3x + 10y + 6z \leqslant 70$$

The crops need fertilizer. One unit of sugar uses 7 units of fertilizer; one unit of rice uses 14 units of fertilizer and a unit of groundnuts is least demanding, using only 4 units. There are 55 fertilizer units available, so the fourth inequality is given by

$$7x + 14y + 4z \leqslant 55$$

The last restriction on crop production in Primonia is the extent of cultivatable land, of which there is a maximum of 50m acres. The land used in growing one unit of the three crops is 5m, 4m and 5m acres respectively, so

$$5x + 4y + 5z \leqslant 50$$

A subjective assessment of the agricultural economy of Primonia might tend towards the conclusion that rice (y) was a demanding crop, especially in terms of transport, fertilizer and capital. An economist, if given the profit margins on each unit of crop, would be likely to suggest that rice was totally uneconomic; the profit function is:

$$P = 6x + y + 5z$$

Sugar is slightly more profitable than groundnuts, but would it be worth devoting all 50m acres to sugar? If not, how many units of groundnuts should be grown? Is it sensible not to grow any rice at all? These questions are virtually impossible to handle subjectively, for there are five inequalities and one profit function to handle simultaneously. The linear programming algorithm can be used, however, as the following calculations show.

(A) Lay out the table of coefficients, constraints, and profits as in Step (iii) of the algorithm. (Steps (i) and (ii) have already been carried out.) The table is

3	6	4	1	0	0	0	0	48
11	6	12	0	1	0	0	0	136
3	10	6	0	0	1	0	0	70
⑦	14	4	0	0	0	1	0	55
5	4	5	0	0	0	0	1	50
−6	−1	−5	0	0	0	0	0	0

(B) Step (iv) involves the selection of the largest negative element in the final row of the table. This is −6 in column 1.

(C) Now, find the results of dividing the numbers in the last column (col. 9) by those in column 1. (The answers are: 16, 12.364, 20.333, 7.857 and 10.) The smallest positive value is 7.857 on row 4, so the element (4,1) —i.e. on row 4, column 1—is the pivot element. It is circled in the table shown at Step (A).

(D) Step (vii) involves transformation of the table according to the rules set out there.

(D1) Divide the elements on the pivot row by the pivot element to give $p_1 = 1$, $p_2 = 2$, $p_3 = 0.571$, $p_4 = p_5 = p_6 = 0$, $p_7 = 0.143$, $p_8 = 0.143$ and $p_9 = 7.857$.

(D2) Use formula (1) to transform all other elements in the table that lie neither on the pivot row (row 4) or the pivot column (column 1). Thus, the elements of the transformed row 1 are

$$r_1^* = r_2 - r_1.p_1 = 3 - 3 \times 1 = 0$$
$$r_2^* = r_2 - r_1.p_2 = 6 - 3 \times 2 = 0$$
$$r_3^* = r_3 - r_1.p_3 = 4 - 3 \times 0.571 = 2.286$$
$$r_4^* = r_4 - r_1.p_4 = 1 - 3 \times 0 = 1$$
$$r_5^* = r_5 - 5_1.p_5 = 0 - 3 \times 0 = 0$$
$$r_6^* = r_6 - r_1.p_6 = 0$$
$$r_7^* = r_7 - r_1.p_7 = 0$$
$$r_8^* = r_8 - r_1.p_8 = 0$$
and $r_9^* = r_9 - r_1.p_9 = 24.429$

Transformed values for the elements of the other non-pivot rows are found by repeated application of Formula (1) to give a new table:

0	0	2.286	1	0	0	−0.429	24.429
0	−16	5.714	0	1	0	−1.571	49.571
0	4	4.286	0	0	1	−0.429	46.429
1	2	0.571	0	0	0	0.143	7.857
0	−6	2.143	0	0	0	−0.714	10.714
0	11	−1.571	0	0	0	0.857	47.714

The profit derived from growing 7.857 units of sugar (1 in row 4 of column 1, 7.857 in row 4 of column 9) is $47.714m. This is not the maximum attainable profit, as there is still a negative element on the last row—in column 3. Thus, a second iteration is necessary. It will be left for the reader to show that the transformed table at the end of the second iteration is:

0	6.4	0	1	0	0	0.333	−1.067	13
0	0.0	0	0	1	0	0.333	−2.667	21
0	16.0	0	0	0	1	1	−2	25
1	3.6	0	0	0	0	0.333	−0.267	5
0	−2.8	1	0	0	0	−0.333	0.467	5
0	6.6	0	0	0	0	0.333	0.733	55

This is an optimal solution, because none of the elements in the last row is negative. The table is interpreted by looking at the first three columns (because there are three crops) and at the last column. The 1 in column 1 lies in row 4. The corresponding element in column 9 is 5, so 5 units of crop 1 (sugar) are grown. Column 2 does not contain a 1, so no rice is grown. 5 units of groundnuts are grown (1 in row 5 of column 3; 5 in row 5 of column 9). Therefore, to achieve maximum profit of $55m, the agricultural land of Primonia should be devoted entirely to the growing of sugar and groundnuts. Rice cultivation is uneconomic, and should be abandoned.

Looking at the five constraints in turn, it can be calculated that $45m out of the $48m available capital is used; 135,000 out of the maximum 136,000 man-hours; only 45 out of the total 70 transport units, but the whole of the fertilizer allocation and the available land.

The algorithm just described is called the Simplex Method. It is very similar to an algorithm developed by Gauss in the nineteenth century for solving sets of linear simultaneous equations. Given the outline of the algorithm (steps (i) to (vii) above) and the two examples, it should be easy to write a computer program to implement the method. Before this is done, one difficulty must be cleared up. Tables 1 to 4 are two-dimensional arrays of numbers, but up to this point only one-dimensional arrays have been used. It would be possible to treat each of the columns (or the rows) of the tables as a one-dimensional array, but this would mean declaring a number of one-dimensional arrays in the DIMENSION statement. One of the advantages of using arrays is that they cut down the number of identifiers (names) that are required, so in the same way that one-dimensional arrays replace individual elements, so do two-dimensional arrays replace individual one-dimensional arrays. The principle can, in theory, be extended to three, four and higher-dimensional arrays but in practice the great majority of programs require only one or two-dimensional arrays.

Two dimensional arrays must be declared in a DIMENSION statement, just as one-dimensional arrays are. The DIMENSION statement for a one-dimensional array specified the maximum number of elements in that array. In the two-dimensional case, the maximum number of rows and columns must be specified. The statement

```
DIMENSION X(20,20),IR(10,5)
```

specifies (i) a two-dimensional array of floating point numbers called X which can hold a maximum of 20 rows and 20 columns—a total of 400 elements—and (ii) a two-dimensional integer array, IR, with a maximum of 10 rows and 5 columns. Note carefully that the maximum number of rows is declared first. (The rule is RC—rows then columns.) The arrays X and IR are shown pictorially in Figure 4.8, though they are not actually held in this form in the store. The position of any element in a two-dimensional

Figure 4.8

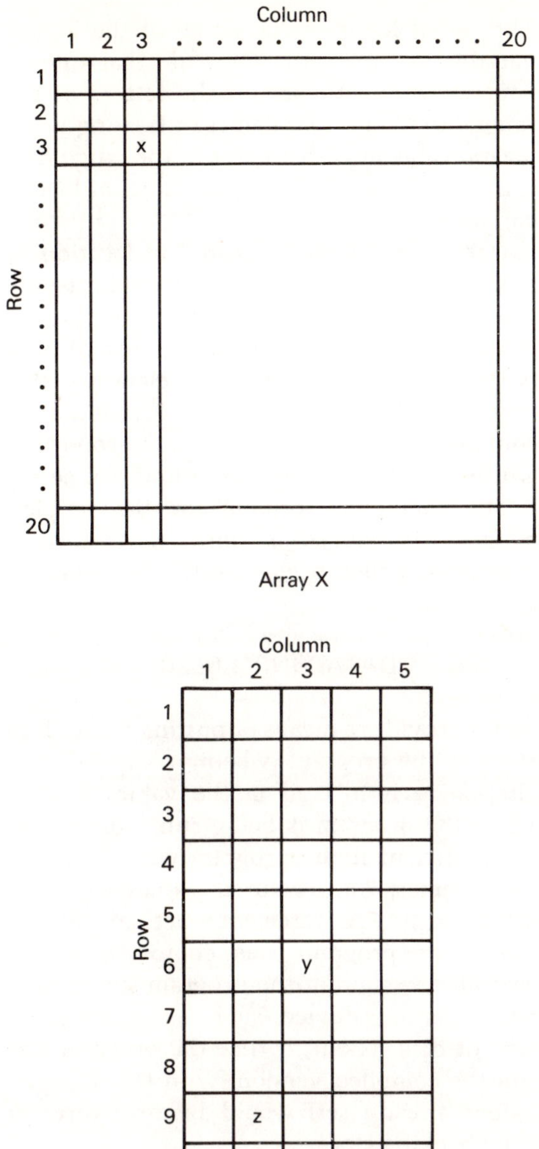

array can be specified exactly by giving the array name followed by the row number then the column number in brackets. For example, X(3,3) is marked by an x in Figure 4.8, while IR(6,3) and IR(9,2) are marked by y and z in the same Figure. In the Linear Programming example, each of the four tables could be represented in Fortran by a two-dimensional array. The DIMENSION statement would set the maximum number of rows (variables, = crops) and columns (= constraints) that the program could handle. Obviously, a smaller number of rows and columns than the number actually declared could be used without modifying the program.

The use of subroutines with two-dimensional arrays as arguments is straightforward. However, the first dimensions of the two-dimensional array in the subroutine must be exactly the same as the first dimensions of the corresponding arrays in the calling program. The second dimensions can be varied in the manner already described, as long as the array in the subroutine has no more columns than the corresponding array in the calling program. Even this restriction can be minimized. Look through the following program to find how both dimensions of a two-dimensional array can be set up at run-time:

```
DIMENSION X(10,5),Y(100,2)
NDIM1 = 10
NDIM2 = 100
N1 = 5
N2 = 2
CALL INOUT(X,NDIM1,N1)
CALL INOUT(Y,NDIM2,N2)
.........................
.........................
STOP
END
SUBROUTINE INOUT(A,N,M)
DIMENSION A(N,M)
.........................
RETURN
END
```

The Simplex algorithm is an iterative one, in that steps (ii) to (vii) must be executed repeatedly until a condition is satisfied; the condition is: there are no negative elements in the last row of the table. This condition is either satisfied or not satisfied—it is a yes/no sort of condition. It could be handled by using an integer indicator which could be set to 1 if the condition were satisfied, and to 0 if it were unsatisfied. (A floating point indicator should not be used, because floating-point numbers are not held exactly in a computer; the quantity 1.0 might be represented as 0.9999999 in a PDP-11 or similar computer working to 7 significant decimal digits. If the floating point location Z was used to hold this indicator, the condition IF(X.EQ.1.0) may never be satisfied, for X may never equal 1.0 exactly, whereas if the integer location IND was used to hold the indicator then the condition IF(IND.EQ.1) could be satisfied, for integers are stored exactly.)

The use of an integer indicator is often perfectly satisfactory. However, a special kind of Fortran variable exists specifically for use in the situation in which a yes/no value is expected. It is called a LOGICAL variable, and can take the values .TRUE. or .FALSE. (the full stops are part of the value). For example, if L is a logical variable, the following statements would be possible:

```
L=.TRUE.
L=.FALSE.
```

Also, one can test whether L contains the value .TRUE. by using an ordinary IF statement of the type:

```
IF(L) GO TO 10
IF(L) STOP
```

Alternatively, a test of whether L contains the value .FALSE. is achieved by

```
IF (.NOT.L) X =X*0.5
IF (.NOT.L) L =.TRUE.
```

If a location is to be used to store logical variables, then the names of those locations must be declared in a special statement which precedes the DIMENSION statement (if any) at the head of the program. The statement is:

```
LOGICAL name 1, name 2, ......, name n
```

For example, a program using the logical variables stored in locations L1 and L2 would begin with the statement

```
LOGICAL L1,L2
```

The initial letter of the name of the logical variable is irrelevant. The rule that integer names begin with one of the letters in the range I to M and that names of floating-point locations begin with a letter in the range A to H or O to Z does not apply to names of logical variables.

One logical variable, stored in location T, will be used in the program to indicate the presence or absence of negative elements in the last row of the table at the end of Step (vii). If there are any negative elements, this variable will take the value .FALSE.. If it contains the value .TRUE. then the condition will be satisfied and the iterations can cease.

The computer program for the simplex method introduces a third new Fortran statement, which occurs on the third line of the program. It is

```
DATA XINF/1.0E20/
```

and it provides a means of putting values into store as the program is being compiled. The alternative is to read in the values as data when the program is being run. This would be inefficient if the program was to be run many times. The required values could be set by a DATA statement once and for all when the program was compiled, and a compiled version of the program stored on a backing-store device such as a magnetic tape or disc. The next time the program was run the compiled version (with the required values already set) would be read directly from backing store.

DATA statements can be used to set up numerical values or alphanumeric (Hollerith) strings. For example,

```
DATA A,B,C/1.0,2.0,3.0/
```

sets the value 1.0 in location, 2.0 in location B, and 3.0 in location C. The statement

```
DATA AD,BC,BP,QR/4*0.0/
```

sets the contents of locations AD, BC, BP and QR equal to 0.0, while the statement

```
DATA AB,CD/4HSTOP,4HFINI/
```

stores the alphanumeric string STOP in location AB and alphanumeric string FINI in location CD. DATA statements can be used to initialize the values stored in a complete array; the two statements

```
DIMENSION X(100),IA(10,15)
DATA X,IA/100*0.0,150*0/
```

sets up two arrays—one (X) a floating point one-dimensional array with 100 elements, and the other (IA) a two-dimensional array with 150 elements. The 100 elements of X are each set to 0.0, and the 150 elements of IA are set to 0 by the DATA statement. It should be noted that this use of DATA statements is not, strictly speaking, standard Fortran; however, most Fortran compilers will accept this deviation from the standard. Otherwise, each element of the array must be independently set to a value:

```
DATA X(1),X(2),. . . .,X(100),IA(1),IA(2),
. . . .,IA(150)/100*0.0,150*0/
```

The values set in DATA statements can be of three forms:

(i) simple constants, such as 10.6, or multiples of constants, for example 100*10.6;

(ii) constants expressed in exponential form, such as 1.0E6, or multiples of such constants. The value 1.0E6 means: 1.0×10^6.

(iii) Hollerith constants, such as 4HABCD. Normally you should not attempt to put more than 4 characters into each location if your program is to run on different makes of computer. Also, some computers will accept the following:

```
DIMENSION X(6)
DATA X/22HREMEMBERΔTOΔ↔
↔SWITCHΔOFF/
```

The 22 characters will be split up into groups containing the maximum number of characters that can be held in one location (4 for PDP-11s, 8 for ICL 1900 series, 10 for CDC-6600 and 7600 computers) and the unused locations in array X will be untouched. Most computers, however, will only accept Hollerith constants of the following form:

```
DIMENSION X(6)
DATA X/4HREME,4HMBER,4HΔTOΔ,↔
↔4HSWIT,4HCHΔO,4HFFΔΔ/
```

The last point to note is that the DATA statement must follow the DIMENSION statement (if there is one).

A Fortran program for the Simplex method is given in Table 4.10. There are ample comments in the listing which should be sufficient to allow you to follow the working of the program without much difficulty, if you also follow the outline of the computing algorithm given above.

EXERCISES

(46) The Utopian water-board is currently investigating possible sources of water-supply for the new town development. The three possible sources are recycled water (x), supplies pumped from a nearby aquifer (y) and water from an existing direct-supply reservoir (z). The gross profit from the sale of each of the three types of water is estimated to be 1, 4 and 5 thousand Utopian dollars respectively per million gallons supplied. There are, however, restrictions on the supply of water, as follows:

(a) the maximum supply of water has been fixed at 100m gallons. The maximum levels of extraction from the three sources are in the proportion: 1 : 2 : 1;

(b) transportation costs must be limited to $120m. The proportional transport costs are: 0 : 2 : 3;

(c) capital costs must not exceed $80m. The proportional unit costs are 2 : 1 : 0 (since the direct-supply reservoir has already been built).

(*continued on p.* 87)

Table 4.10

```
      LOGICAL T
      DIMENSION V(11,30)
C THE FOLLOWING DATA STATEMENT PUTS THE VALUE OF TEN TO THE POWER TWENTY
C INTO LOCATION XINF
      DATA XINF/1.0E20/
C THE FIRST 20 ELEMENTS OF COL. 1 OF ARRAY V ARE USED TO HOLD AN
C ALPHANUMERIC TITLE, WHICH IS READ AND PRINTED IMMEDIATELY.  THE LATER
C USE OF ARRAY V TO HOLD NUMBERS IS NOT AFFECTED.
      READ(5,152) (V(I,1),I=1,20)
  152 FORMAT(20A4)
      WRITE(6,153) (V(I,1),I=1,20)
  153 FORMAT(12HOJOB NAME = ,20A4)
      READ(5,1000) NV, NC
 1000 FORMAT(2I3)
C NV = NO OF REAL VARIABLES
C NC = NO OF INEQUALITIES
      NOO=NC+1
C NOO IS NUMBER OF ROWS IN TABLE (NC INEQUALITIES + PROFIT FUNCTION)
      NCONST=NC+NV
C NO = NUMBER OF COLUMNS IN TABLE
      NO=NCONST+1
C LOP IS 1 MORE THAN THE NUMBER OF COLUMNS.  COLUMN LOP OF THE TABLE
C IS USED AT STEP 5 TO HOLD THE DIVIDENDS.
      LOP=NO+1
      WRITE(6,66)
   66 FORMAT(22HOINITIAL SIMPLEX TABLE//)
C READ TABLE INTO ARRAY V
      DO 1 I=1,NOO
      READ(5,1001) (V(I,J),J=1,NO)
 1001 FORMAT(8F10.0)
    1 WRITE(6,102) (V(I,J),J=1,NO)
  102 FORMAT(1H ,15F8.3)
C
C DATA IN ORDER:
C CARD 1:JOB NAME.  USE UP TO 80 COLUMNS.
C CARD 2: COLS 1-3 - NV;  COLS 4-6 - NC.
C CARD 3 (AND SUBSEQUENT CARDS AS NECESSARY):
C PUNCH THE ROWS OF THE SIMPLEX TABLE STARTING EACH ROW
C ON A NEW CARD IN (8F10.0) FORMAT.  PUT THE 'REAL'
C VARIABLES FIRST, THEN THE SLACK VARIABLES, THEN THE VALUE OF THE
C CONSTRAINT.  THE LAST ROW OF ALL CONTAINS THE NEGATED COEFFICIENTS
C OF THE PROFIT FUNCTION WITH A ZERO FOR EACH SLACK VARIABLE AND FOR
C THE CONSTRAINT.  FOR EXAMPLE IF NV=3 AND NC=5 THERE WILL BE (5+3+1)=8
C COLUMNS IN THE SIMPLEX TABLEAU AND (5+1)=6 ROWS.
      DO 2 ITER=1,100
C LOOP 2 IS THE MAJOR LOOP, EXECUTED ONCE PER ITERATION OF THE SIMPLEX
```

Table 4.10 (continued)

```
C ALGORITHM
C AN UPPER LIMIT OF 100 ITERATIONS HAS BEEN SET ARBITRARILY.
      T=.TRUE.
      S=0.0
C LOOP 3 SEARCHES FOR NEGATIVE ELEMNTS IN THE BOTTOM ROW OF THE TABLE
      DO 3 I=1,NCONST
      IF(V(NOO,I).GE.S) GOTO 3
C IF A NEGATIVE ELEMENT, PUT COLUMN NUMBER IN K AND SET T TO .FALSE.
      K=I
      T=.FALSE.
      S=V(NOO,I)
    3 CONTINUE
C IF T IS STILL TRUE THEN EXIT FROM LOOP 2 AND GOTO LABEL 75
      IF(T) GOTO 75
      DO 4 I=1,NOO
C LOOP 4 CARRIES OUT STEP 5.  NOTE EXTRA COLUMN OF TABLE IS USED TO
C STORE RESULTS OF DIVISION OF ELEMENTS OF LAST COLUMN BY ELEMENTS OF
C COLUMN K.  TO PREVENT DIVISION BY ZERO, A CHECK IS MADE ON THE VALUE
C IN ROW I OF COLUMN K.  IF IT IS ZERO THE DIVISION IS NOT CARRIED OUT,
C AND A VERY LARGE VALUE IS PUT INTO ROW I OF COLUMN LOP.
C (DIVISION BY ZERO CAUSES A FATAL ERROR USUALLY TERMED 'OVERFLOW SET')
      IF(V(I,K).NE.0) GOTO 40
      V(I,LOP)=XINF
      GOTO 4
   40 V(I,LOP)=V(I,NO)/V(I,K)
    4 CONTINUE
      S=V(1,LOP)
      KK=1
      DO 5 I=2,NOO
C LOOP 5: LOOK FOR SMALLEST POSITIVE VALUE IN COLUMN LOP.  STORE ITS
C ROW NUMBER IN KK.
      IF(V(I,LOP).GT.S.OR.V(I,LOP).LE.0.0) GOTO 5
      S=V(I,LOP)
      KK=I
    5 CONTINUE
C KK IS THE INDEX OF THE PIVOT ROW
C PIVOT ELEMENT IS V(KK,K).  (STEP 6)
      PIV=V(KK,K)
C NOW BEGIN STEP 7 - TRANSFORMATION OF THE TABLE.  FIRSTLY, DIVIDE
C THE PIVOT ROW BY THE PIVOT ELEMENT TO GIVE NEW PIVOT ROW (P IN TEXT)
      DO 6 I=1,NO
    6 V(KK,I)=V(KK,I)/PIV
C NEXT USE FORMULA (1) TO TRANSFORM ELEMENTS NOT ON THE PIVOT ROW
      DO 8 I=1,NOO
C OMIT THE PIVOT ROW - IT HAS ALREADY BEEN TRANSFORMED.
      IF(I.EQ.KK) GOTO 8
```

Table 4.10 (continued)

```
      RK=V(I,K)
      DO 9 J=1,NO
    9 V(I,J)=V(I,J)-RK*V(KK,J)
    8 CONTINUE
C PRINT OUT REVISED TABLE
      WRITE(6,456)
  456 FORMAT(14HOREVISED TABLE//)
      DO 452 I=1,NOO
  452 WRITE(6,102) (V(I,J),J=1,NO)
C END OF LOOP 2. IF LESS THAN 100 ITERATIONS, REPEAT SIMPLEX CYCLE
    2 CONTINUE
C IF PROGRAM REACHES THIS POINT NO SOLUTION HAS BEEN FOUND.
      WRITE(6,139)
  139 FORMAT(18HONO SOLUTION FOUND)
      STOP
C LOOP 75 IS REACHED ONLY WHEN NO NEGATIVE ELEMENTS ARE FOUND ON LAST
C ROW OF THE TABLE IN LOOP 3 ABOVE.
   75 WRITE(6,221)
  221 FORMAT(20HOFINAL SIMPLEX TABLE//)
      DO 30 I=1,NOO
   30 WRITE(6,102) (V(I,J),J=1,NO)
      WRITE(6,201) V(NOO,NO)
  201 FORMAT(10HOPROFIT = ,F12.5)
      WRITE(6,202) ITER
  202 FORMAT(1HO,I6,11H ITERATIONS)
C LOOP 20 SEARCHES THE FIRST NV COLUMNS TO FIND THOSE STARTING WITH
C A 0 OR A 1.  NOTICE THE USE OF INT(V(1,I)+0.0000001).  THIS IS DONE TO
C ELIMINATE THE EFFECTS OF ROUNDING ERROR - THE RESULT OF A FLOATING-
C POINT OPERATION MAY BE 1.0 EXACTLY, BUT THIS RESULT COULD BE STORED
C AS 0.9999999 OR 1.0000001. NEITHER OF THESE IS EQUAL TO 1.0, SO THE
C 'IF' STATEMENT WOULD NOT BE  SATISFIED. HOWEVER, IF 0.0000001 IS
C ADDED AND THE INTEGER PART TAKEN THEN THE RESULT WILL BE STORED AS
C 1 EXACTLY, BECAUSE INTEGERS ARE NEITHER ROUNDED NOR TRUNCATED.
      DO 20 I=1,NV
      K=INT(V(1,I)+0.0000001)
      IF(K.NE.0.AND.K.NE.1) GOTO 23
C IF COLUMN STARTS WITH 0 OR 1, FIND THE ROW NUMBER OF THE 1 AND STORE
C IT IN LOCATION K.
      NOO1=NOO-1
      DO 21 J=1,NOO1
      IF(INT(V(J,I)+0.0000001).NE.1) GOTO 21
      K=J
      GOTO 22
   21 CONTINUE
C PRINT VALUE IN ROW K OF LAST COLUMN
   22 WRITE(6,210) I,V(K,NO)
```

Table 4.10 (continued)

```
  210 FORMAT(19H VALUE FOR VARIABLE,I4,4H  IS,F10.4)
C IF COLUMN DOES NOT BEGIN WITH 0 OR 1 THEN NO UNITS OF THAT PARTICULAR
C VARIABLE ARE PRODUCED
      GOTO 20
   23 Z=0.
      WRITE(6,210) I,Z
   20 CONTINUE
      STOP
      END
```

How much water should the water board extract from each of the three sources?

(47) Write a Fortran program to print histograms on the lineprinter. Use the following strategy.

(i) Set up an array IX of dimension 20 to hold the number of items in each class. (If more than 20 classes are to be used increase the dimension of IX.) Also declare an array SLINE of dimension 50.

(ii) Put the characters '*' and 'Δ' (blank) into locations STAR and BLANK respectively, using a DATA statement. Also, fill SLINE with 50 blank characters, also using a DATA statement.

(iii) Read the number of classes (N) and the number of items in each class (IX(I),I = 1,N).

(iv) Convert the contents of array IX to numbers in the range 1–50 by the formula

$$x_i* = ((x_i - x_{min})/(\text{range of } x))*49.0 + 1$$

where x_i* is the ith transformed value, x_i the ith actual value, and x_{min} the smallest observed actual value. The range of x is $x_{max} - x_{min}$. (Note: this must be done in floating point. Why?)

(v) For each of the N classes:
 (a) find K = IX(I);
 (b) fill the first K elements of SLINE with the character STAR (by the use of a DO loop and the statement SLINE(I) = STAR);
 (c) print out the first K elements of SLINE using (50A1) format;
 (d) fill the first K elements of SLINE with the character BLANK.

Now use this program to produce a histogram from the following data, which refer to the amount spent (in pence) by a random sample of 47 people questioned in a local shopping centre.

class	0–20	20–50	50–100	over 100
frequency	10	7	12	18

4.6 Linear Regression Analysis

Regression analysis is one of the most widely-used statistical methods. Gregory (1963), Cole and King (1968) and Ebdon (1976) give illustrations of its applications in geography. Readers are referred to one of these texts for an introduction to the method, which is basically concerned with summarizing the relationship between two variables, an explanatory or independent variable and a dependent variable. The explanatory variable will be denoted by the letter x and the dependent variable by the letter y. If a given set of measurements is a random sample from a defined population then statistical significance tests can be applied to test the null hypothesis that the relationship observed in the sample does not exist in the population as a whole, that is, the sample relationship is due entirely to chance.

A regression relationship is usually linear, being defined by the equation

$$y = \alpha + \beta x + \varepsilon$$

Any equation of this form is linear, though the

graphical representation of the functional relationship may not necessarily be a straight line. Thus, the equation $y = \alpha + \beta x^2 + \varepsilon$ does not plot as a straight line, though it is a linear equation. In this section we are more concerned with the computational aspects of regression so it will be assumed that both the equation and the corresponding graph show linear relationships. In such cases, the parameter β represents the slope of the regression line (which is the line defined by the regression equation) and the other parameter α gives the point at which the regression line (extended if necessary) cuts the vertical (y) axis. These two parameters are written β and α when a population is being studied; if they are sample estimators then caps are added to indicate this, so $\hat{\beta}$ is an estimator of β and $\hat{\alpha}$ is an estimator of α. From this point onwards it will be assumed that we are dealing with sample estimators.

The regression line is calculated so that the sum of the squared distances from observed points to the points on the regression line that are immediately above or below them is minimized. That is why the regression line is often termed the 'least-squares line'. The difference between an observed point and the point vertically above or below it on the regression line is called a residual, and the residual for the ith point is written e_i. Figure 4.9 shows five points, labelled p_1 to p_5, which are located at (x_1,y_1), (x_2,y_2), ..., (x_5,y_5). The regression line runs through this scatter of points so that the sum of squares of the

Figure 4.9

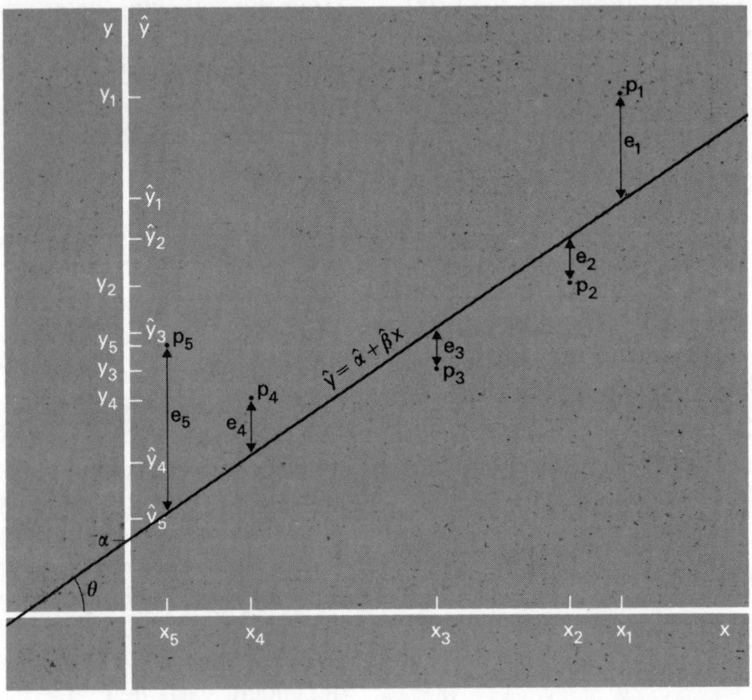

(vertical) deviations from the points to the line is minimized. The vertical axis is the y axis, hence the deviations are the differences between the y coordinates of the points and the y coordinates of the points immediately above or below them on the regression line. These latter points have coordinates $(x_1,\hat{y}_1)$, $(x_2,\hat{y}_2)$, ..., $(x_5,\hat{y}_5)$ and the ŷs are called *predicted values* of y.

The calculations involved in finding $\hat{\alpha}$ and $\hat{\beta}$ are not complicated. Given a set of n pairs of observed values (x_i,y_i) the slope parameter is defined by:

$$\hat{\beta}=\frac{\sum_{i=1}^{n}(x_i-\bar{x})^2(y_i-\bar{y})^2}{\sum_{i=1}^{n}(x_i-\bar{x})^2}$$

and

$$\hat{\alpha}=\bar{y}-\hat{\beta}\bar{x}$$

where x and y are the means of x and y respectively.

Since it is assumed that the (x_i,y_i) are a random sample, it is possible to ask questions such as:

(i) What is the interval in which the population values α and β are likely to lie?

(ii) Given the sample regression line $\hat{y}=\hat{\alpha}+\hat{\beta}x$ what is the interval in which the true regression line is likely to lie?

(iii) What is the probability that a value of $\hat{\beta}$ as high as the value computed could have been derived from a random sample drawn from a population in which the true value of β is actually zero?

To answer these equations interval estimates of α, β and y are needed. In the calculation of these interval estimates, or confidence intervals, the following values are required:
(a) the variance of the residuals, written s^2 for a sample; this is found from

$$s^2=\sum_{i=1}^{n}e_i^2/(n-2)$$

(b) the standard errors of $\hat{\alpha}$ and $\hat{\beta}$, written $s_{\hat{\alpha}}$ and $s_{\hat{\beta}}$ These values are found from:

$$s_{\hat{\alpha}}=\left\{s^2\left[\frac{1}{n}+\frac{\bar{x}^2}{\sum_{i=1}^{n}(x_i-\bar{x})^2}\right]\right\}^{\frac{1}{2}}$$

$$s_{\hat{\beta}}=\left[\frac{s^2}{\sum_{i=1}^{n}(x_i-\bar{x})^2}\right]^{\frac{1}{2}}$$

(c) the tabled value of Student's t with $(n-2)$ degrees of freedom and significance level p if the $100(1-p)$ percent confidence limits are required. For example, if 95% confidence intervals are required then $100(1-p)=95$ and $p=0.05$. Thus, if n is 12 and $p=0.05$ then $t_{10,0.05}$ is 2.228.

Once these quantities are available the answers to the three questions posed above can be given, as follows:

(i) $100(1-p)$ percent confidence limits for α and β are found from the equations

$$\hat{\alpha}-s_{\hat{\alpha}}.t_{n-2,p}\leqslant\alpha\leqslant\hat{\alpha}+s_{\hat{\alpha}}.t_{n-2,p}$$
$$\hat{\beta}-s_{\hat{\beta}}.t_{n-2,p}\leqslant\beta\leqslant\hat{\beta}+s_{\hat{\beta}}.t_{n-2,p}$$

In other words, if the random sampling process was repeated a large number of times then $100(1-p)$ percent of the computed confidence intervals would contain the true values of α and β.

(ii) $100(1-p)$ percent confidence intervals for the population regression line $(\alpha+\beta x)$ are calculated at each of n given data points from:

$$ylow\leqslant(\alpha+\beta x)\leqslant yhi$$

where

$$ylow=\hat{y}_i-band,$$
$$yhi\ =\hat{y}_i+band,$$
$$band=t_{n-2,p}.s_{\hat{y}},$$

and

$$s_{\hat{y}}=\left\{s^2\left[\frac{1}{n}+\frac{(x_i-\bar{x})^2}{\sum_{i=1}^{n}(x_i-\bar{x})^2}\right]\right\}^{\frac{1}{2}}$$

(iii) The null hypothesis $\beta=0$ is tested by the F statistic against the alternative hypotheses $\beta>0$ and $\beta<0$. The value of F is

$$F=SSR/(SSE/(n-2))$$

where

$$SSR=\hat{\beta}^2\sum_{i=1}^{n}(x_i-\bar{x})^2$$

and

$$SSE=\sum_{i=1}^{n}(y_i-\bar{y})^2-SSR$$

If the calculated F exceeds the tabled value of F at significance level (p/2) with 1 and (n – 2) degrees of freedom then the null hypothesis is not accepted at significance level p. (The F table is entered with significance level (p/2) and not p because the test is two-tailed; if a one-tailed test was used, with alternative hypothesis $\beta < 0$ or $\beta > 0$, then enter the F table with significance level p.) As an alternative to the F ratio you can use $t = \sqrt{F}$ with (n – 2) degrees of freedom and significance level as before.

Table 4.11

Cost of journey x	*Number of visits* y
100	55
90	70
80	90
70	100
70	90
70	105
70	80
65	110
60	125
60	115
55	130
50	130

The coefficient of determination, r^2, gives the proportion of the variance of y that is accounted for by the regression. It can be calculated from the values already found:

$$r^2 = SSR/\sum_{i=1}^{n} (y_i - \bar{y})^2$$

As an example of the use of regression analysis, consider the data matrix shown in Table 4.11. The dependent variable (y) is the number of people out of a sample of 840 who were recorded as visiting a certain town to shop. The explanatory variable, x, is the cost of the journey to and from their town of origin. The results of the analysis, using Example Program 16 (Table 4.13), are shown in Table 4.12.

Table 4.12

```
REGRESSION  EQUATION  FOR  THIS  DATA IS Y =  210. 444 +   -1. 578X

      X          Y      PREDICTED Y    RESIDUAL

  100. 000    55. 000      52. 667       2. 333
   90. 000    70. 000      68. 444       1. 556
   80. 000    90. 000      84. 222       5. 778
   70. 000   100. 000     100. 000       0. 000
   70. 000    90. 000     100. 000     -10. 000
   70. 000   105. 000     100. 000       5. 000
   70. 000    80. 000     100. 000     -20. 000
   65. 000   110. 000     107. 889       2. 111
   60. 000   125. 000     115. 778       9. 222
   60. 000   115. 000     115. 778      -0. 778
   55. 000   130. 000     123. 667       6. 333
   50. 000   130. 000     131. 556      -1. 556

EST.  STANDARD ERROR OF ALPHA =   12. 571
EST.  STANDARD ERROR OF BETA  =    0. 176
VARIANCE OF RESIDUALS =   69. 889

  95. OPERCENT CONFIDENCE LIMITS FOR ALPHA AND BETA ARE

  182. 437   ALPHA     238. 452
```

Table 4.12 (continued)

```
 -1. 970   BETA      -1. 185

  95. OPERCENT CONFIDENCE LIMITS FOR PREDICTED Y ARE
```

	X	Y	YBAR	LOWER	UPPER
1	100. 0000	55. 0000	52. 6667	39. 7175	65. 6159
2	90. 0000	70. 0000	68. 4444	58. 9267	77. 9621
3	80. 0000	90. 0000	84. 2222	77. 5642	90. 8803
4	70. 0000	100. 0000	100. 0000	94. 6231	105. 3769
5	70. 0000	90. 0000	100. 0000	94. 6231	105. 3769
6	70. 0000	105. 0000	100. 0000	94. 6231	105. 3769
7	70. 0000	80. 0000	100. 0000	94. 6231	105. 3769
8	65. 0000	110. 0000	107. 8889	102. 1648	113. 6130
9	60. 0000	125. 0000	115. 7778	109. 1197	122. 4358
10	60. 0000	115. 0000	115. 7778	109. 1197	122. 4358
11	55. 0000	130. 0000	123. 6667	115. 6915	131. 6418
12	50. 0000	130. 0000	131. 5556	122. 0379	141. 0733

```
TOTAL SUM OF SQUARES =      6300. 0000
REGRESSION (EXPLAINED) SUM OF SQUARES =      5601. 1108
CORRELATION COEFFICIENT =  0. 943
COEFFICIENT OF DETERMINATION =  88. 907 PERCENT

TEST OF HO: BETA=0 IN POPULATION
(ALTERNATIVE: BETA>0 OR BETA<0)
F =      80. 1431 WITH 1 AND  10 DEGREES OF FREEDOM

LOOK UP F TABLES AT SIGNIFICANCE LEVEL (P/2) SINCE THIS IS A 2-TAILED TEST

ALTERNATIVELY CHECK T =       8. 9523 WITH  10 DEGREES OF FREEDOM AGAINST
                           TABLED VALUE OF T AT SIGNIFICANCE LEVEL (P/2)

P IN BOTH CASES IS THE PROBABILITY LEVEL, E. G. 5 PERCENT

IF TABLED VALUE EXCEEDS PRINTED VALUE, ACCEPT NULL HYPOTHESIS (HO)
```

Table 4.13

```
C $$$$$$$$$$$$$$$$$$$$$$$$$$$$$$$$$$$$$$$$$$$$$*****************************$$$$$$$$$$$$$$$$
C                  EXAMPLE PROGRAM 16
C THIS PROGRAM CALCULATES THE LINEAR REGRESSION OF Y ON X AND ASSOCIATED
C STATISTICS AND CONFIDENCE LIMITS.
C DATA IS SUPPLIED AS FOLLOWS
C CARD 1 N THE NUMBER OF OBSERVATIONS IN (I3) FORMAT
```

Table 4.13 (continued)

```
C CARD 2 (CONTINUING ON TO NEW CARDS AS NECCESSARY) VALUES OF X IN
C FORMAT(20F4.0). THIS FORMAT CAN BE EASILY ALTERED BY MODIFYING
C THE FORMAT STATEMENT LABELLED 2.
C CARD 3 DATA FOR Y IN THE SAME ORDER AND THE SAME FORMAT AS THE X DATA.
C CARD 4 COLUMNS 1-10 TVALUE, THE TABLED VALUE OF T AT THE APPROPRIATE
C SIGNIFICANCE LEVEL WITH (N-2) DEGREES OF FREEDOM. FOR EXAMPLE,
C THE TVALUE FOR 95% PROBABILITY AND 10 DEGREES OF FREEDOM
C IS 2.228
C COLUMNS 11-20 PROB. THE PROBABILITY LEVEL (EG. 95.0 IN THE
C PREVIOUS EXAMPLE)
C THESE LAST TWO ITEMS ARE PUNCHED IN (2F10.0) FORMAT
C $$$$$$$$$$$$$$$$$$$$$$$$$$$$$$$$$$$$$$$$$$$$$$$$$$$$$$$$$$$$$$$$$$$$$$$$$$$
      DIMENSION X(100),Y(100),YCAP(100),E(100)
C
C THESE DIMENSIONS MUST BE INCREASED IF THE NUMBER OF CASES EXCEEDS 100
C
      WRITE(6,1111)
 1111 FORMAT(25H1SIMPLE LINEAR REGRESSION/22H0PROGRAM BY P.M. MATHER/
     1 39H0GEOGRAPHY DEPT., NOTTINGHAM UNIVERSITY/10H0JULY,1974)
      READ(5,1) N
    1 FORMAT(I3)
C
C READ X AND Y VALUES
C
      READ(5,2) (X(I),I=1,N)
      READ(5,2) (Y(I),I=1,N)
    2 FORMAT(20F4.0)
      XMEAN=0.0
      YMEAN=0.0
      FN=N
C
C CALCULATE MEANS OF X AND Y
C
      DO 3 I=1,N
      XMEAN=XMEAN+X(I)
    3 YMEAN=YMEAN+Y(I)
      XMEAN=XMEAN/FN
      YMEAN=YMEAN/FN
C
C CALCULATE SUM OF PRODUCTS OF DEVIATIONS OF X AND Y FROM THEIR
C RESPECTIVE MEANS IN SIGXY, ALSO SUM OF SQUARES OF DEVIATIONS OF X FROM
C THE MEAN OF X IN SIGX2, AND SUM OF SQUARES OF THE Y FROM THE MEAN
C OF Y IN SIGY2.
C NOTE THAT  SIGY2  IS THE 'TOTAL SUM OF SQUARES CORRECTED FOR THE MEAN'
C
      SIGXY=0.0
```

Table 4.13 (continued)

```
      SIGX2=0.0
      SIGY2=0.0
      DO 4 I=1,N
C
C Z IS DEVIATION OF CURRENT X(I) FROM MEAN OF X
C ZZ IS DEVIATION OF CURRENT Y(I) FROM MEAN OF Y.
C SIGX2 IS ACCUMULATING SQUARES OF DEVIATIONS OF X
C SIGY2 IS ACCUMULATING SQUARES OF DEVIATIONS OF Y
C SIGXY IS ACCUMULATING PRODUCT OF DEVIATIONS OF X AND Y
C
      Z=X(I)-XMEAN
      ZZ=Y(I)-YMEAN
      SIGY2=SIGY2+ZZ*ZZ
      SIGX2=SIGX2+Z*Z
    4 SIGXY=SIGXY+Z*ZZ
C
C REGRESSION ESTIMATES
C
      BETA=SIGXY/SIGX2
C BETA IS SLOPE PARAMETER
      ALPHA=YMEAN-BETA*XMEAN
C ALPHA IS INTERCEPT
      WRITE(6,5) ALPHA , BETA
    5 FORMAT(46HOREGRESSION  EQUATION  FOR  THIS  DATA IS Y = ,F8.3,
     13H +,F8.3,1HX//1H ,6X,1HX,10X,1HY,4X,11HPREDICTED Y,3X,
     28HRESIDUAL/)
      S=0.0
C
C PREDICTED VALUES OF Y AND RESIDUALS
C
      DO 6 I=1,N
      YCAP(I)=ALPHA+BETA*X(I)
      E(I)=Y(I)-YCAP(I)
C YCAP IS PREDICTED VALUE
C E IS RESIDUAL
C S IS SUM OF SQUARED RESIDUALS
      S=S+E(I)**2
    6 WRITE(6,7) X(I), Y(I), YCAP(I), E(I)
    7 FORMAT(1H ,2F10.3,2F12.3)
C
C VARIANCE OF RESIDUALS
C
      S=S/FLOAT(N-2)
C
C STANDARD ERRORS
C
```

Table 4.13 (continued)

```
      SEALFA=SQRT(S*(1.0/FN+XMEAN**2/SIGX2))
      SEBETA=SQRT(S/SIGX2)
      WRITE(6,8) SEALFA, SEBETA, S
    8 FORMAT(32HOEST.  STANDARD ERROR OF ALPHA = ,F8.3/
     1       32H EST.  STANDARD ERROR OF BETA   = ,F8.3/
     2       25H VARIANCE OF RESIDUALS = ,F8.3)
      NDF=N-2
      READ(5,9) TVALUE, PROB
    9 FORMAT(2F10.0)
C
C CONFIDENCE INTERVALS AROUND ALPHA AND BETA
C
  XY=TVALUE*SEALFA
      ALIM1=ALPHA-XY
      ALIM2=ALPHA+XY
      XY=TVALUE*SEBETA
      BLIM1=BETA-XY
      BLIM2=BETA+XY
      WRITE(6,10) PROB, ALIM1, ALIM2, BLIM1, BLIM2
   10 FORMAT(1HO,F6.1,48HPERCENT CONFFDENCE LIMITS FOR ALPHA AND BETA AR
     1E/1HO,F9.3,2X,5HALPHA,2X,F9.3/1HO,F9.3,2X,4HBETA,3X,F9.3)
      WRITE(6,11) PROB
   11 FORMAT(1HO,F6.1,45HPERCENT CONFIDENCE LIMITS FOR PREDICTED Y ARE/
     1 1HO,9X,1HX,9X,1HY,7X,4HYBAR,7X,5HLOWER,4X,5HUPPER)
C
C CONFIDENCE LIMITS AROUND YBAR
C
      DO 12 I=1,N
      VARY=S*(1.0/FN+((X(I)-XMEAN)**2)/SIGX2)
      BAND=TVALUE*SQRT(VARY)
      Y1=YCAP(I)-BAND
      Y2=YCAP(I)+BAND
   12 WRITE(6,13) I, X(I), Y(I), YCAP(I), Y1, Y2
   13 FORMAT(1H ,I4,5F10.4)
C
C NEXT, CALCULATE SUMMARY STATISTICS
C
      SSR=BETA*BETA*SIGX2
      RSQ=SSR/SIGY2
      R=SQRT(RSQ)
      RSQ=RSQ*100.0
      WRITE(6,14) SIGY2, SSR, R, RSQ
   14 FORMAT(24HOTOTAL SUM OF SQUARES = ,F12.4/41H REGRESSION (EXPLAINED
     1) SUM OF SQUARES = ,F12.4/27H CORRELATION COEFFICIENT = ,F6.3/
     2 32H COEFFICIENT OF DETERMINATION = ,F7.3,8H PERCENT)
```

Table 4.13 (continued)

```
C TEST OF NULL HYPOTHESIS B=0
C
      SSE=SIGY2-SSR
      F=SSR/(SSE/FLOAT(NDF))
      WRITE(6,15) F, NDF
   15 FORMAT(33HOTEST OF HO: BETA=0 IN POPULATION/32H (ALTERNATIVE: BETA
     1>0 OR BETA<0)/5H F = ,F10.4,11H WITH 1 AND,I4,19H DEGREES OF FREED
     2OM)
      T=SQRT(F)
      WRITE(6,16) T, NDF
   16 FORMAT(75HOLOOK UP F TABLES AT SIGNIFICANCE LEVEL (P/2) SINCE THIS
     1 IS A 2-TAILED TEST/25HOALTERNATIVELY CHECK T = ,F10.4,5H WITH,I4,
     273H DEGREES OF FREEDOM AGAINST TABLED VALUE OF T AT SIGNIFICANCE L
     3EVEL (P/2)/56HOP IN BOTH CASES IS THE PROBABILITY LEVEL, E.G.5 PER
     4CENT/67HOIF TABLED VALUE EXCEEDS PRINTED VALUE, ACCEPT NULL HYPOTH
     5ESIS (HO))
      STOP
      END
```

EXERCISES

(48) Write a brief report, including appropriate diagrams, on the results of the regression analysis presented in Table 4.12.

(49) Modify Example Program 16 so that it calls a subroutine HIST which plots a histogram of the frequency distribution of the residuals. You could modify the program you wrote as Exercise (47) to serve this purpose.

(50) Add an option to Example Program 16 to allow for the transformation of either the x values, the y values, or both. You will need to write a subroutine to perform the transformations, and read in a LOGICAL variable to determine whether or not the transformation routine is to be called, using L1 format. If it is called it should read an integer variable which indicates the transformation to be applied. Allow for the following transformations:

1. logarithm
2. square
3. cube
4. square root.

Use the modified program to find the sample regression equation and associated confidence intervals and significance tests for the following data, which consists of values of stream width (y) and peak annual discharge (x) for 24 drainage basins. Take the logarithm of the peak annual discharge.

x	y	x	y	x	y
5.7	63.0	12.0	93.0	13.0	35.0
17.0	260.0	19.0	210.0	19.0	156.0
22.0	92.0	24.0	130.0	26.0	120.0
31.0	230.0	39.0	163.0	43.0	230.0
50.0	720.0	52.0	740.0	60.0	2000.0
61.0	890.0	70.0	2900.0	80.0	1300.0
85.0	2500.0	90.0	840.0	90.0	2500.0
120.0	1150.0	130.0	5100.0	170.0	5600.0

(51) The binomial probability law (L. J. King, 1969, page 40) is given by:

$$\text{Prob}(x) = \frac{N!}{x!(N-x)!} p^x q^{(N-x)}$$

N is the number of observations, p the probability of outcome 1 and q the probability of outcome 2. The symbol N! means $N \times (N-1) \times (N-2) \times \ldots\ldots \times 1$. For example, consider the probability of getting x = 0,1,2,3,4,5, 6,7,8,9,10 heads in an experiment which consists of tossing an unbiased coin 10 times. In this case, N = 10, and p = q = 0.5. Thus, the probability of the outcome being 5 heads and 5 tails is:

$$\text{Prob}(x=5) = \frac{10!}{5!\,5!} \times 0.5^5 \times 0.5^5 = 0.2460.$$

Write a Fortran program to evaluate the probabilities of the binomial distribution for any N, p and q and for a specified range of x values. Remember that the xs must be whole numbers, and that 0! = 1.

(52) The Normal probability law is:

$$\text{Prob}(x) = \frac{1}{\sigma\sqrt{2\pi}}.\exp[-\tfrac{1}{2}(x-m)^2/\sigma^2]$$

The symbol 'exp' means 'e raised to the power of'. The value e is a special number in mathematics, and is equal to 2.7183. The actual value of e need not be known, because there is a Fortran library function (like SQRT and ALOG) which returns the value of e^x, where x is a floating point argument. The function name is EXP, so to evaluate e^x simply write: Y = EXP(X). The parameters m and σ^2 are the mean and variance of the distribution.

Write a Fortran program to calculate the value of Prob(x) for up to 30 x values. Read in the values of m and σ^2. Provide explanatory captions. (Note: set π (pi) equal to 3.141592653.)

(53) A number of methods are available to test whether a given sequence is random or not. One of the simplest is the 'runs test', which is best explained by example. Imagine that you have collected information on the occupants of the houses along a particular road, and that you are interested in finding out the probability that the households possessing colour televisions are located randomly along the road. If n, the total number of houses, is 45 and if the possession of a colour television is indicated by a + sign, with the absence of a colour television being shown by a − sign, the following sequence might result:

\+ + − − + − − − + − + + − + + + + + − +
\+ + + + + − − − + − − − + + + − − − −
− + + + + + − +

Firstly, count the number of + and − signs (n_1 and n_2 respectively). Now count the number of runs, defined as the number of groups of successive + or − signs plus the number of + and − signs occurring alone. Call the number of runs r. In the example, $n_1 = 25$, $n_2 = 20$ and r = 19. If both n_1 and n_2 are less than 20 then look up tables which give the probability of as few as (or as many as) r runs occurring in a sequence of n_1 pluses and n_2 minuses. If either n_1 or n_2 exceeds 20 a z transformation can be used. The value of z is calculated from:

$$z = \frac{r - \left(\dfrac{2n_1n_2}{n_1+n_2} + 1\right)}{\sqrt{\left\{\dfrac{2n_1n_2(2n_1n_2 - n_1 - n_2)}{(n_1+n_2)^2(n_1+n_2-1)}\right\}}}$$

For the values of n_1, n_2 and r given above z is 1.2897. The probability associated with the value 1.2897 in tables of the standard normal distribution is about 0.097. This exceeds the traditional 5% significance point (p = 0.05) so the null hypothesis that the observed distribution could have arisen as a result of chance is accepted at p = 0.05.

Write a Fortran program to carry out the procedure described above, using the example data as a test.

APPENDIX A

Operation of IBM Model 29 Card Punch; Fortran Coding Conventions

A.1 Use of IBM Model 29 Card Punch

The normal form of input for both programs and data is the punch card. A brief account of the method of operation of one of the most widely used card punches, the IBM Model 29, is given here. Other punches have a similar mode of operation.

An IBM Model 29 punch was shown in Figure 2.1. A more detailed photograph of the keyboard is shown in Figure A.1. The steps to be followed in operating the punch are:

(i) Ensure punch is plugged in, and switch on (the on/off switch is below and to the right of the desk top on which the keyboard rests. Note that up is 'on' and down is 'off').

(ii) A supply of fresh cards should be placed in the input hopper, which is at the top right (looking from the front) behind the IBM badge. The cards should be placed in the hopper with the printed side facing the user, and row 9 at the bottom.

(iii) Turn the five leftmost switches on the back sloping plate of the keyboard to the 'up' position. The rightmost switch (marked CLEAR) will not normally stay in the 'up' position.

(iv) Press the key marked FEED twice. This will bring two cards from the input hopper to the punching station.

(v) You can now punch the top card by pressing the appropriate keys. The NUMERIC key is equivalent to the upper-case key on a typewriter. Other special keys are present, but are not normally used. If the keyboard jams, press the ERROR RESET key which is immediately above the NUMERIC key.

(vi) On completion of punching the first card, press the REL key. This does two things: (a) moves the card that has just been punched, and (b) brings another card down from the input hopper.

(vii) Repeat steps (v) and (vi) until punching is completed. Now lift up the CLEAR key at the right hand side of the sloping plate at the back of the keyboard, and hold it in the 'up' position for a second or two. The cards still in the punching station will be moved along to the output hopper. You can now remove the cards from the output hopper and switch off the machine.

The basic operations are thus quite simple. More complicated uses of the machine are described in the Reference Manual for the Model 29 punch, which is published by IBM Systems Development Corporation, New York 12602. The reference number of the manual is GA24-3332-5.

To avoid the possibility of unnecessary errors, never leave mispunched cards lying around the punch room. They may be inadvertently included in someone's job. Check

Figure A.1

your cards carefully before submitting them. It is bad practice to submit jobs that have not been visually checked for punching errors.

A.2 Fortran Coding Conventions

Fortran programs are normally punched on cards such as the one shown in Figure A.2. The program cards, preceded by the job control cards and followed by the data cards, have to be punched according to a set of rules. It is important to remember that these rules refer only to the *program* cards, and not to the job control cards (which usually start in column 1, but see your local Computing Centre documentation) or to data cards (the layout of which is controlled by the input format specifications as outlined in Section 3.2). The rules for program cards are:

(i) Columns 1 to 5 inclusive are reserved for numerical labels such as the labels of FORMAT statements (Section 3.2) or the terminal labels of DO loops (Section 3.4). Such labels can be punched anywhere within columns 1 to 5, but it is visually more pleasing if the labels are punched according to a consistent rule, for example, always starting in column 1, or

Figure A.2

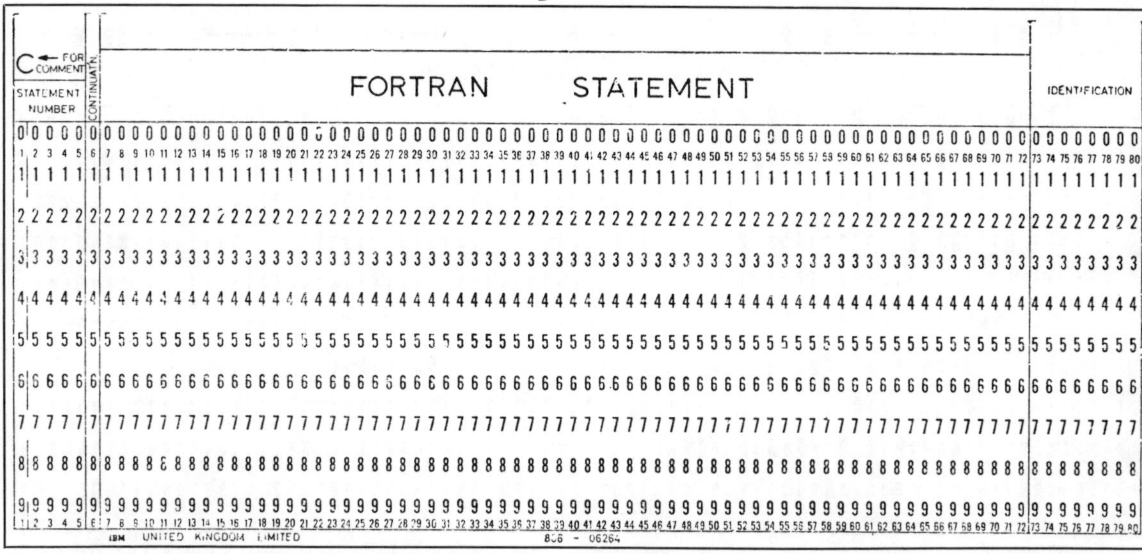

else putting units in column 5, tens in column 4 and so on. Remember that the same label can be used once only within a program or subroutine. Labels need not follow any mathematical sequence, but some programmers ensure that labels start at 1 and proceed in sequence. This makes the program logic easier to follow, but is not essential.

(ii) Column 6 is normally left blank. If any character other than zero is punched in this column then the instruction contained on the card is treated as a continuation of the instruction on the preceding card. For example:

```
101    FORMAT(25H0THISΔISΔANΔEXAMPLEΔOFΔAΔ,
      *18HΔCONTINUATIONΔCARD//)
```

Some computer systems limit the number of continuation cards in one sequence, but this

Figure A.3

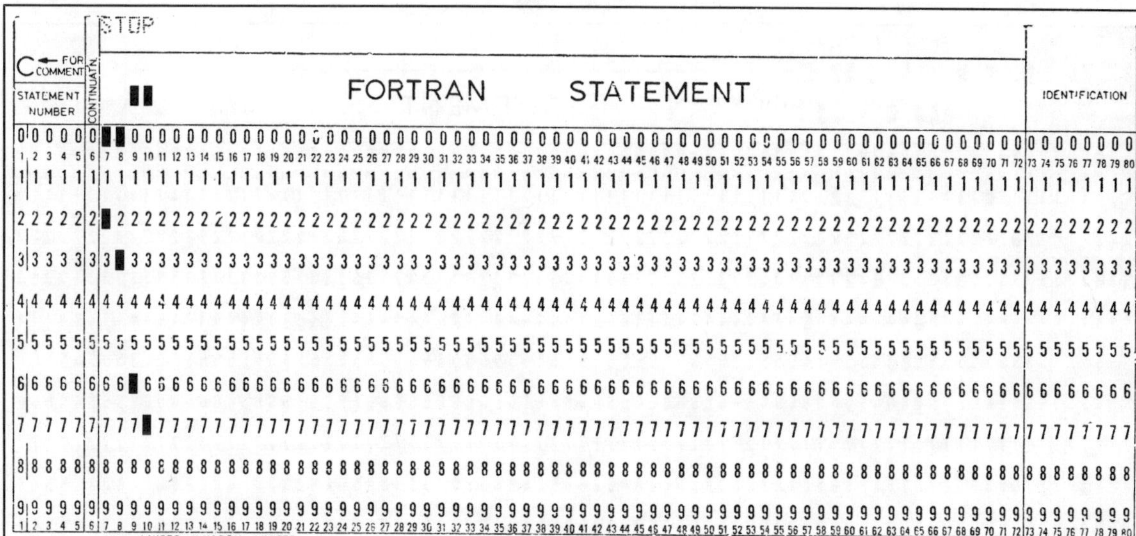

Figure A.4

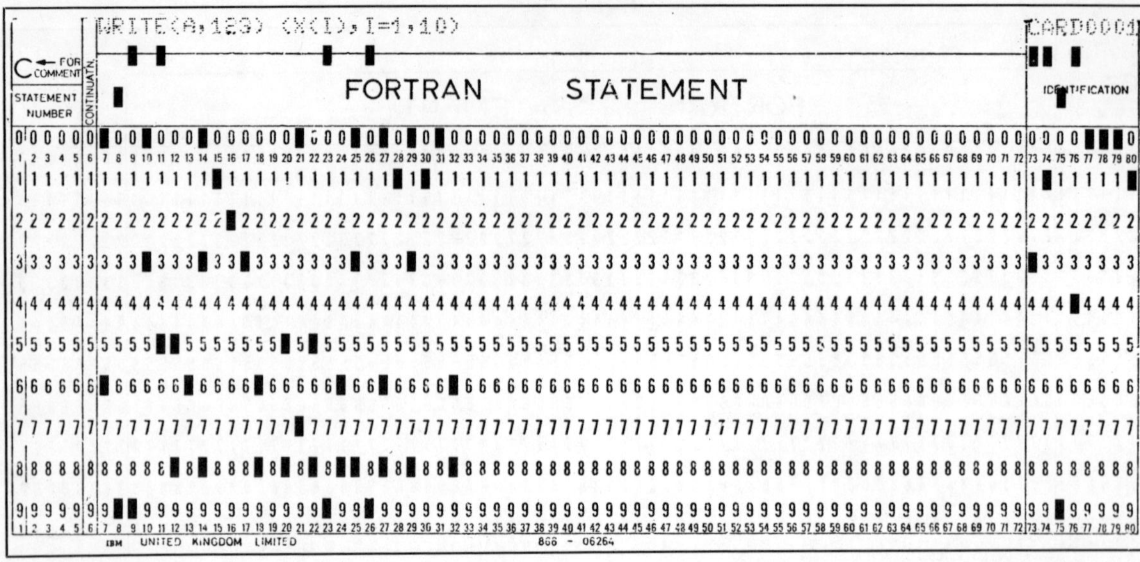

is rarely likely to inconvenience most users.

(iii) Columns 7 to 72 inclusive contain the Fortran instruction. Only one instruction is allowed on one card. If more space is needed, use a continuation card or cards. An example of a card containing Fortran instructions is shown in Figure A.3.

(iv) Columns 73 to 80 can be used to give the card an alphanumeric identifier (Figure A.4). Alternatively, these columns can be left blank. As it is not unknown for cards to get out of order, numbering or otherwise labelling the program cards in columns 73 to 80 could save a lot of time and trouble.

(v) Column 1 is normally within the label field, as described in paragraph (i) of this section. If the letter C is punched in column 1 then the entire contents of the card will be ignored by the compiler. This is a useful way of indicating what the program is doing, for the contents of the card—although ignored by the compiler—are printed in the program listing. An example of a comment card is shown in Figure A.5.

Figure A.5

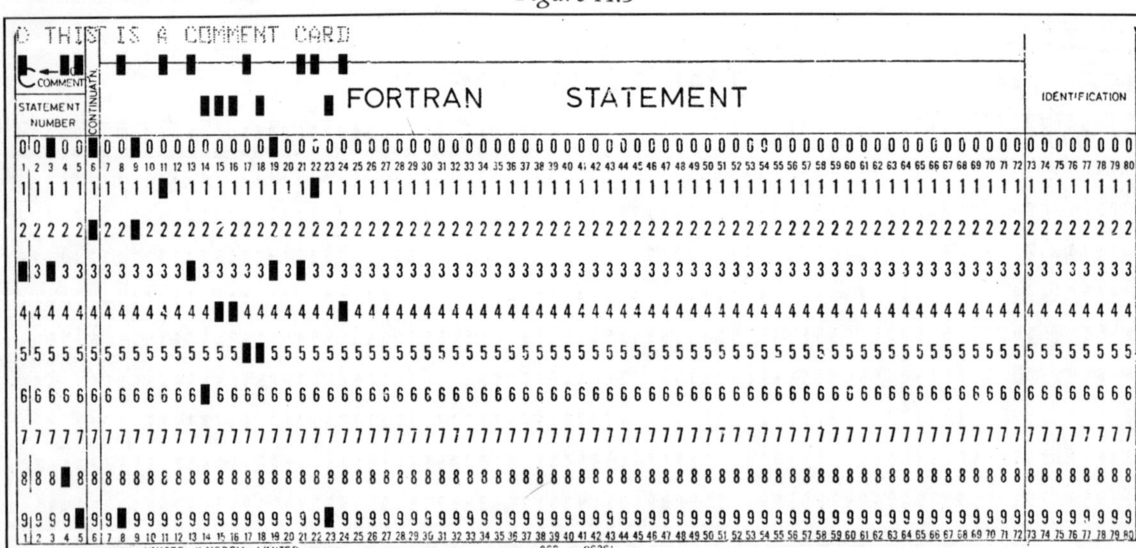

APPENDIX B

Sources of Fortran Programs

Useful Fortran programs for a wide variety of geographical problems can be obtained from many university and college computing centres and from individual research workers. Other sources are books and journals, some of which are given below.

(1) *Computer Applications in the Natural and Social Sciences,* Department of Geography, The University, Nottingham NG7 2RD. Mainly Fortran programs for multivariate statistical methods, computer mapping and data analysis.

(2) *Computer Contributions,* State Geological Survey, The University of Kansas, Lawrence, Kansas 66044, USA. Mainly Fortran programs for data analysis, computer mapping, simulation, etc.

(3) Geographic Program Exchange (GPE) is an organization which distributes programs of geographical interest. Details are available from: Dr. R. I. Wittick, Computer Institute for Social Science Research, Michigan State University, East Lansing, Michigan 48823, USA.

(4) L. Baker, *A Selection of Geographical Computer Programs,* Geographical Papers No. 6, London School of Economics and Political Science, Houghton Street, Aldwych, London.

(5) D. F. Marble, *Some Computer Programs for Geographical Research,* Department of Geography, Northwestern University, Evanston, Illinois.

(6) W. R. Tobler, *Selected Computer Programs,* Department of Geography, University of Michigan, Ann Arbor, Michigan.

The following books contain computer programs of general interest as well as descriptions of their application:

(7) W. W. Cooley and P. R. Lohnes, *Multivariate Data Analysis,* John Wiley & Sons, Inc., New York, 1971. Mainly advanced statistical programs for principal components analysis, discriminant analysis and analysis of variance.

(8) P. R. Lohnes and W. W. Cooley, *Introduction to Statistical Procedures with Computer Exercises,* John Wiley & Sons, Inc., New York, 1968. Contains programs for statistical tests and exercises. Covers random number generation and simulation of statistical frequency distributions.

(9) J. W. Harbaugh and G. Bonham-Carter, *Computer Simulation in Geology,* John Wiley & Sons, Inc., New York, 1970. Mainly of interest to physical geographers, but includes discussion of linear programming and random number generation as well as a number of simulation programs.

(10) R. E. Blackith and R. A. Reyment, *Multivariate Morphometrics,* Academic Press, New York, 1971. Contains Fortran programs for multivariate data analysis.

(11) J. C. Davies, *Statistics and Data Analysis*

in Geology, John Wiley & Sons, Inc., New York, 1973. Lists many Fortran programs, ranging from the elementary (e.g. mean and standard deviation) to the advanced (e.g. factor analysis).

(12) M. R. Anderberg, *Cluster Analysis for Applications*, Academic Press, New York, 1974. Programs for every conceivable form of cluster analysis, but make sure you get the errata slip.

(13) P. M. Mather, *Computational Methods of Multivariate Analysis in Physical Geography*, John Wiley & Sons, Ltd., Chichester, forthcoming, 1976. Fortran programs for multivariate analysis, classification and discrimination. Also an extensive subroutine library.

The following journals publish Fortran programs and subroutines which may be of interest to geographers:

(14) *Journal of the Association of Mathematical Geology*.

(15) *Communications of the Association of Computer Manufacturers (CACM)*.

(16) *Computer Journal*.

(17) *Journal of the Royal Statistical Society, Series C (Applied Statistics)*.

(18) *Geocom Bulletin*.

APPENDIX C

Error Detection

The most successful way of avoiding errors in programming is to think carefully about the computing algorithm before punching any cards. The case studies in Chapter 4 were intended to show that no amount of programming ability will compensate for a lack of awareness of the nature of the problem to be programmed. However, even the most careful computer user will make errors in punching out his program; these errors may range from simple punching errors, such as omitting one or more characters, or punching the wrong characters, to the complete omission of several lines of program. Of course, logical errors are not uncommon, for example, failure to realize that under some circumstances a particular instruction may cause the computer to attempt to divide by zero.

No system of checking can be entirely foolproof. If the name of a store location is misspelled it is unlikely that the checking system would detect the error unless one of the Fortran naming conventions had been violated (e.g. the mis-spelled name may begin with a numeral, or may contain an illegal character) in which case the checking system would detect an error. If you punch the word DALAM when you really meant to punch DLAM then the compiler (which does the checking) will simply assume that you are using two separate floating point locations, DALAM and DLAM.

Logical errors can be detected only if they result in instructions which cause the computer to attempt to execute some impossible operation. Division by zero is not logically possible, so any Fortran instruction which involves division by zero will be detected. However, if you meant to evaluate $\sum_{i=1}^{n} x_i^2$ but calculated $(\sum_{i=1}^{n} x_i)^2$ instead, the program would run and would give an answer, but not the correct one. The only way to check a program for logical errors of this nature is to test the program using data which produce an answer which is known to be correct. Never assume a program is working just because it runs without any errors being discovered.

A large class of errors can be detected automatically by the Fortran compiler, or by the operating system. These errors are of three kinds:

(a) those that result from the violation of one or more of the rules of Fortran;

(b) those that result from instructions that the computer is not capable of executing; and

(c) those resulting from faulty job control cards.

Type (a) are known as 'compile–time' errors, and type (b) are known as 'execution errors' or 'run–time' errors. Type (c) are job control errors.

Table C.1

```
0001              DIMENSIOM X(30)

[M X(30)]
ERROR   21
W UNRECOGNIZED STATEMENT.

0002              S=0.0
0003              DO 1 I=1,10
0004            1 X(I)=FLOAT(I)
                  DO 2 I=1,10

[(I)=FLOAT(]
ERROR   0
I REDUNDANT CONTINUATION MARK; IT IS IGNORED
[FLOAT(I)O ]
ERROR   29
W ILLEGAL CHAR. TERMINATING A STMT. OR POSS. BAD OPERATOR.
[(I)=FLOAT(]
ERROR   21
W UNRECOGNIZED STATEMENT.

0005            2 S=S+X(I)
                  WRITE(6,100) S
0006
0007          100 FIRMAT(1HO,F12.4)

[,F12.4)]
ERROR   21
W UNRECOGNIZED STATEMENT.

0008    STOP

[]
ERROR   2
I ILLEGAL STMT. NUMBER, NON-NUMERIC CHAR. IN COLS. 1-5
[]
ERROR   21
W UNRECOGNIZED STATEMENT.

0009              END

[END]
ERROR  114
W MISSING DO LOOP TERMINATION(S).
```

(a) COMPILE–TIME ERRORS

Every Fortran compiler checks for numerous possible errors in the programs it is translating. Each error has an associated error number, though the error numbers differ from one compiler to another. Lists of error numbers and their meanings can be found in the manufacturers' handbooks and in computing centre literature. For example, error number 57 in the ICL 1900 Fortran compiler means that the same label has been used twice within one program unit (i.e. the same main program or subroutine). Usually the error number and its meaning are printed after the offending statement, or else at the end of the program unit. One real error can generate several spurious errors. A missing FORMAT statement that is referenced by several READ or WRITE statements will result in one error message for each reference. An example of spurious compile—time errors is shown in Table C.1. The first error, 'unrecognized statement', is due to the fact that the word DIMENSION is mis-spelled. The next three error messages ('redundant continuation mark', 'illegal character terminating statement', and 'unrecognized statement') are all produced by the same error—the instructions X(I) = FLOAT(I) and DO 2 I = 1,10 are begun in column 6 instead of column 7. The first error message, 'redundant continuation mark', reveals the true nature of the error, while the other two are spurious. A further error message at the end of the program ('missing DO loop termination(s)') is due to the fact that the line labelled 1, which terminates the DO loop beginning on the third line of the program, was not interpreted because the X of X(I) was punched in column 6 instead of column 7. The other error messages are straightforward—the word FORMAT is mis-spelled on line number 7, and punching STOP in columns 1–4 rather than 7–10 on line 8. Because of the compile–time errors the job is not executed and a message such as 'job abandoned' or 'execution deleted' is printed.

A second example is given in Table C.2. Most of the errors are self-evident, except possibly the one on the sixth line where an attempt was made to link two IF statements

```
IF(X.GT.0.0) STOP
IF(X.LT. -6.8) STOP
```

by means of an .OR. instruction. If this is done, the resulting instruction should take the form:

```
IF(X.GT.0.0.OR.X.LT. -6.8) STOP
```

The instruction .AND. can be used in a similar way.

The example shown in Table C.2 shows that some compilers fail to detect even gross errors. On the seventh line of the program the statement WRITE(5,12) is specified. This is illogical as the number 5 refers to the card reader. A more subtle error is the instruction GOTO 5 on the ninth line. Label 5 is the terminal statement of a DO loop. One of the rules for using DO loops is that you can jump (GOTO) to a label outside the loop when control is inside the loop, but you cannot jump (GOTO) to a label inside a loop when control is outside the loop. Even though neither of these errors was detected at compilation stage, it is likely that they would be picked up during execution.

(b) EXECUTION ERRORS

The fact that a program has compiled without error does not mean that it will run correctly, as the example given in Table C.2 shows. The program listed in Table C.3 has also compiled correctly, but the instruction on the third line (Z = Y/X) cannot be executed because X is zero. Computers cannot divide by zero. A message ('real division by zero') is printed and the job is stopped ('aborted'). The word 'real' in the error message means 'floating point'. Usually, the line on which the error occurred is given.

The next example, Table C.4, shows two errors, both of which result from an attempt

Table C.2

```
0001          1 FORMAT(8F0.0

[T(8F0.0]
ERROR   59
W TOO MANY LEFT PARENTHESES IN FORMAT.

0002            READ(5,1) X
0003            13-I

[13-I]
ERROR   21
W UNRECOGNIZED STATEMENT.

0004            DO 5 I=I=1,17

[DO 5 I=I=1]
ERROR   29
W ILLEGAL CHAR. TERMINATING A STMT. OR POSS. BAD OPERATOR.

0005          5 X=X/2.0
0006            IF(X.GT.0.0.OR.LT.-6.8) STOP

[8) STOP]
ERROR   21
W UNRECOGNIZED STATEMENT.

0007            WRITE(5,12)
0008         12 FORMAT(15HOPASSED LABEL 5)
0009            GOTO 5
0010            STOP

[STOP]
ERROR  109
I NO PATH TO THIS STATEMENT.

0011            END
```

Table C.3

```
0001            X=0.0
0002            Y=10.0
0003            Z=Y/X
0004            WRITE(6,10) Z
0005         10 FORMAT(1H0,F10.3)
0006            STOP
0007            END
```

Table C.3 (continued)

```
FORT003008 REAL DIVISION BY ZERO
NAME     SEQ
MAIN.    00003
```

Table C.4

```
      DIMENSION X(10)
      N=11
      DO 1 I=1,N
    1 X(I)=I
      WRITE(6,2) (X(I),I=1,N)
    2 FORMAT(1H0,10F10.2)
      STOP
      END
```

```
FORT007001 SUBSCRIPT GREATER THAN DIMENSIONED
NAME     SEQ
MAIN.    00004
```

```
11.00      2.00      3.00      4.00      5.00      6.00      7.00↔
                                        ↔8.00      9.00     10.00

11.00
```

to store or retrieve 11 numbers from an array X which was given a DIMENSION of 10. The DEC Fortran operating system, under which this program was run, is obviously more liberal than other operating systems because it has allowed the program to continue after the first execution error. The results output by the program show that X(1) contains the value 11.0 instead of 1.0—an indication of the unpredictable effects of exceeding the DIMENSION of an array.

Some execution errors are so serious that the program is halted immediately and no diagnostic error message printed. (Some systems manage to produce a cryptic message such as 'reservation violation' or 'illegal instruction'.) Usually such errors result not from an attempt to carry out an undefined instruction, such as division by zero, but from an attempt to execute an unintelligible instruction. Such instructions can be produced by (i) exceeding the dimensions of an array. In the example above, the contents of X(1) were overwritten by the value 11.0. If a program instruction had been overwritten by 11.0 then the operating system would not be able to interpret '11.0' and the program would be halted; (ii) incorrect number of arguments in a CALL statement. The number and type of arguments must agree in the CALL statement and the SUBROUTINE statement; (iii) use of actual values (such as 1.4142) as function or subroutine arguments when the argument is altered within the subroutine or function. This can lead to execution errors or—even worse—to undetectable errors in the program output. Consider the following example:

```
B = FUNCT(2,3,6)
. . . . . . . . . . . . . . . . .
. . . . . . . . . . . . . . . . .
STOP
END
FUNCTION FUNCT(I,J,K)
I = 10
J = 11
K = 12
RETURN
END
```

On return, the computer behaves as if 2 is really 10, 3 is 11 and 6 is 12.

The amount of information printed out after an execution error has occurred depends on what is known as the 'trace level', or degree of control the operating system has over the program. In the examples above, the error message identified the error and the line on which it occurred. On occasions when the cause of a particularly devious error is being sought the values of the variables in use at the time the error occurred may be required. Most systems will, on request, print out these values. The instruction TRACE 2 in ICL 1900 job control will cause the results of a maximum of 70 instructions preceding the error to be printed. The use of these high trace levels results in longer execution times, sometimes of the order of 20 to 30 times, so once a program has been run and tested the lowest trace level should be used. High trace levels are designed for use only in error detection.

(c) JOB CONTROL ERRORS

These are the hardest to describe in a general textbook since the job control language or JCL varies from computer to computer. JCL errors cause the job to be terminated immediately. Error messages may be given, but they are likely to be cryptic. In the example given in Table C.5 the message 'command syntax error' is printed because, in the JCL used in this example, all commands begin with a $ sign. The command &RUN NELLY is therefore incorrect because it begins with a & sign.

Table C.5

```
$JOB NELLY[222,222]
DATE:-01-JAN-75
TIME:-16:38:27

$EXECUTE
FORTRAN   V06.13
#SY:NELLY ,LP:<BI:/GO

0001      DIMENSION X(10)
0002      N=11
0003      DO 1 I=1,N
0004   1  X(I)=I
0005      WRITE(6,2) (X(I),I=1,N)
0006   2  FORMAT(1H0,10F10.2)
0007      STOP
0008      END

#
$RU LINK
LINK V01-04
NELLY<NELLY/CC,FTNLIB/L/E

SPACE USED 010414, SPACE FREE↔
                        ↔021134

*FATAL*-COMMAND SYNTAX ERROR

&RU NELLY

$FINISH
```

APPENDIX D

Answers to Selected Exercises

1.

```
      A=17.2 + 97.63
      WRITE(6,10) A
   10 FORMAT(1H ,F6.2)
      STOP
      END
```

2.

```
      B = 19.4 - 10.1
      WRITE(6,917) B
  917 FORMAT(1H ,F4.1)
      STOP
      END
```

4.

```
      READ(5,107) IA,IB
  107 FORMAT(2I4)
      IPROD=IA*IB
      WRITE(6,91) IPROD
   91 FORMAT(1H ,I8)
      STOP
      END
```

5. (b) X is the name of a floating point location but format 10 specifies an integer (I).
   (c) (i) superfluous comma following K;
       (ii) superfluous comma after 1H1,;
       (iii) I formats have the structure aIb not aIb.c.
   (d) (i) The READ statement specifies that the number is to be placed in an integer location (IJ) while the format specifies a floating point quantity (code F);
       (ii) Even if F were the correct code, the number of decimal places cannot exceed the total number of characters.
       (iii) The characters 1H0, are editing characters for output formats only.
6. (i) F = 6.0 + 5.0
   (ii) R = 7.2 + 3.6 – (2.7/13.9)
   (iii) AREA = 3.1415926*RAD*RAD
   (iv) G = ((A – B)*(A – B)) + ((C – D)*(C – D))
   (v) D = S/A

7.

```
      READ(5,10) Q,CS
   10 FORMAT(2F10.0)
      QS=(Q*CS) / 1000.0
      WRITE(6,20) QS
   20 FORMAT(1H ,F12.4)
      STOP
      END
```

8.

```
      READ(5,90) O1,O2,O3,O4,O5
   90 FORMAT(5F10.0)
      CHI1=((O1-6.0)*(O1-6.0))/6.0
      CHI2=((O2-6.0)*(O2-6.0))/6.0
      CHI3=((O3-6.0)*(O3-6.0))/6.0
      CHI4=((O4-6.0)*(O4-6.0))/6.0
      CHI5=((O5-6.0)*(O5-6.0))/6.0
```

```
    CHISQ=CHI1+CHI2+CHI3+CHI4↔
                         ↔+CHI2
    WRITE(6,101) CHISQ
101 FORMAT(1H ,F10.4)
    STOP
    END
```

9. (iv) Yes—floating point; (v) yes—floating point; (vi) yes—integer; (vii) yes—floating point; (viii) no—too many characters (the maximum is 6); (ix) no—∗ is not a legal character. Names must consist of up to 6 characters chosen from the 26 upper case letters A–Z and the 10 digits 0–9. The initial character must be a letter. (x) yes—floating point.
10. (i) yes; (ii) no—the name does not begin with a letter; (iii) no—zero subscripts are not allowed; (iv) no—the array dimension is 5 so JOHN(10) does not exist; (v) no—subscript is negative; (vi) no—the subscript must be an integer constant or the name of an integer variable.
11. Floating point.

12.

```
    DIMENSION OBS(250)
    READ(5,1) N
  1 FORMAT(I3)
    READ(5,2) (OBS(I),I=1,N)
  2 FORMAT(10F8.0)
    SUM=0.0
    FN=N
    DO 3 I=1,N
  3 SUM=SUM+OBS(I)
    XMEAN=SUM/FN
    WRITE(6,4) XMEAN
  4 FORMAT(1H ,F12.4)
    STOP
    END
```

13. 173388.8888.
14. Because XBAR and VARX are the names of floating point (not integer) locations.
15. 10; each value is punched in an 8-column field.
16. (i) DO loop ending at label 2 sums the X(I); (ii) DO loop ending at label 3 sums the squared deviations of the X(I) from their mean.

17.

```
    DIMENSION X(600)
    READ(5,6) N
  6 FORMAT(I3)
    READ(5,1) (X(I),I=1,N)
  1 FORMAT(8F10.0)
    XBAR=0.0
    VARX=0.0
    DO 2 I=1,N
  2 XBAR=XBAR+X(I)
    XBAR=XBAR/FLOAT(N)
    DO 3 I=1,N
  3 VARX=VARX+(X(I)-XBAR)*(X(I)↔
                        ↔-XBAR)
    VARX=VARX/FLOAT(N)
    WRITE(6,4) XBAR, VARX
  4 FORMAT(1H ,2F10.4)
    STOP
    END
```

19. 18.5500% and 7.4985%; 9.2600% and 1.5044%.
20.

I	X	XNEW	CONV	A
3	26.2401	15.0255	0.7461	100.0
4	15.0255	10.8404	0.3863	100.0

22. XNEW = 0.5∗(X + A/X).
23. Insert before line I = I + 1:

```
      WRITE(6,5) I, XNEW
    5 FORMAT(1HΔ,I4,F11.5).
```

24. See Table D.1.
25. Insert before line STOP:

```
      N2=N-2
      FN2=FLOAT(N2)
      T=(R*SQRT(FN2))/SQRT(1.0-R*R)
      WRITE(6,11) T, N2
   11 FORMAT(5H0TΔ=Δ,F10.4,5HΔ↔
   ↔WITH,*15,19HDEGREESΔOFΔ↔
   ↔FREEDOM)
```

Table D.1

```
CORRELATION COEFFICIENT PROGRAM

JOB NAME EXAMPLE PROGRAM 11   --- TEST OUTPUT

VARIABLE NAMES ARE X = BASIN AREA      AND Y = TOT LNTH STREAMS

LISTING OF INPUT DATA

          X           Y


   1     4.0800     9.4000
   2     4.5300    10.8500
   3     4.9200    10.6900
   4     6.1800    10.9700
   5     6.2200    11.9400
   6     6.5400    18.6900
   7     7.5500    17.2100
   8     8.9200    18.1400
   9    10.1600    20.1400
  10    12.5000    20.6800

CORRELATION BETWEEN BASIN AREA AND TOT LNTH STREAMS IS 0.8784
```

31.

```
COORDINATES FOR  10  POINTS


   1     2.7000     6.8000
   2     6.1000     2.9000
   3     3.7000     0.0000
   4     2.4000     1.6000
   5     5.9000     4.7000
   6     8.1000     1.6000
   7     6.2000     6.1000
   8     7.3000     5.2000
   9     8.1000     2.9000
  10     5.0000     2.7000

NEAREST NEIGHBOUR STATISTIC =   0.4740
```

32.

```
      DIMENSION X1(100),X2(100)
      READ(5,1) N
    1 FORMAT(I3)
      READ(5,2) (X1(I),X2(I),I=1,N)
    2 FORMAT(2F10.0)
      FN=FLOAT(N)
      D=0.0
      XBAR1=0.0
      XBAR2=0.0
      DO 3 I=1,N
      XBAR1=XBAR1+X1(I)
      XBAR2=XBAR2+X2(I)
      DO 3 J=1,N
      DIST=(X1(I)-X1(J))**2+(X2(I)-X2(J))**2
    3 D=D+DIST
      D=SQRT(D/FN**2)
      XBAR1=XBAR1/FN
      XBAR2=XBAR2/FN
      STDIST=0.0
      DO 4 I=1,N
    4 STDIST=STDIST+(X1(I)-XBAR1)**2+(X2(I)-XBAR2)**2
      STDIST=SQRT(2.0*STDIST/FN)
      WRITE(6,5) D, XBAR1, XBAR2, STDIST
    5 FORMAT(27H0MEAN QUADRATIC DISTANCE = ,F10.4/15H MEAN CENTRE = ,
     1F10.4,5H  AND,F10.4/21H STANDARD DISTANCE = ,F10.4)
      STOP
      END
```

39.

```
C SET UP NECESSARY ARRAYS.  IF N IS GREATER THAN 100 INCREASE DIMENSIONS
      DIMENSION X1(100),X2(100),X(100),RANK(100),IWORK(100)
      WRITE(6,100)
  100 FORMAT(24H0DATA MATRIX - X1 AND X2//)
      READ(5,2) END1
C END1 IS THE VALUE PUNCHED ON THE LAST DATA CARD TO INDICATE THE END
C OF THE DATA
C NOW READ VALUES INTO X1 AND X2, IGNORING EQUAL VALUES.  COUNT THE
C NUMBER OF UNEQUAL VALUES IN N
      N=1
    1 READ(5,2) X1(N), X2(N)
C X1 AND X2 ARE ARRAYS HOLDING PAIRED VALUES
C IS THIS THE LAST DATA CARD?
      IF(X1(N).EQ.END1) GOTO 20
```

```
      WRITE(6,110) X1(N), X2(N)
C ARE THE VALUES X1 AND X2 EQUAL?
      IF(ABS(X1(N)-X2(N)).LT.0.0000001) GOTO 1
      N=N+1
      GOTO 1
    2 FORMAT(2F3.0)
C REDUCE N BY 1 AS LAST CARD HAS BEEN COUNTED AS A 'REAL' DATA CARD
   20 N=N-1
  110 FORMAT(1H ,2F10.4)
C FIND ABSOLUTE DIFFERENCES
      DO 3 I=1,N
    3 X(I)=ABS(X1(I)-X2(I))
C RANK DIFFERENCES FROM LOW TO HIGH
      I=0
    4 I=I+1
      SMALL=1000000.0
      DO 5 J=1,N
      IF(X(J).GE.SMALL) GOTO 5
      K=J
      SMALL=X(J)
    5 CONTINUE
C CHECK FOR EQUAL SCORES AND FIND AVERAGE RANK
      NO=0
      DO 40 J=1,N
      IF(ABS(X(J)-SMALL).GT.0.0000001) GOTO 40
      NO=NO+1
      IWORK(NO)=J
   40 CONTINUE
      SUM=FLOAT(I+I+NO-1)/2.0
      DO 41 J=1,NO
      K=IWORK(J)
      RANK(K)=SUM
C IF X1(K)>X2(K) MAKE RANK NEGATIVE
      IF(X1(K).GT.X2(K)) RANK(K)=-RANK(K)
C SET X(K) TO LARGE NUMBER TO AVOID RE-SELECTION
   41 X(K)=1000001.0
      I=I+NO-1
      IF(I.LT.N) GOTO 4
C NOW SUM RANKS FOR X1<X2 AND X1>X2
      SUM1=0.0
      SUM2=0.0
      DO 6 I=1,N
      IF(RANK(I).GT.0.0) SUM1=SUM1+RANK(I)
      IF(RANK(I).LT.0.0) SUM2=SUM2-RANK(I)
C IF YOU ARE PUZZLED BY THE LAST INSTRUCTION, REMEMBER THAT TWO MINUSES
C ARE EQUIVALENT TO A PLUS.
    6 CONTINUE
```

```
C FIND SMALLER OF SUM1 AND SUM2 USING FORTRAN LIBRARY FUNCTION AMIN1
C AMIN1 WILL RETURN THE (FLOATING POINT) VALUE OF THE SMALLEST OF THE
C FLOATING POINT ARGUMENTS SUBMITTED.
      F3=AMIN1(SUM1,SUM2)
C USE FLOATING POINT ARITHMETIC TO AVOID INTEGER OVERFLOW (I.E. TRYING
C TO STORE AN INTEGER LARGER THAN THE CAPACITY OF THE MACHINE.)
      F1=FLOAT(N)
      F2=FLOAT(N+1)
C COMPUTE Z
      Z=(0.25*F1*F2-F3)/SQRT((F1*F2*(F1+F1+1.0))/24.0)
      WRITE(6,7) Z
    7 FORMAT(21H0WILCOXON TEST :  Z = ,F10.4)
      STOP
      END
```

40. First exercise:

```
      DIMENSION FP(25)
      N=25
      READ(5,10) (FP(I),I=1,25)
   10 FORMAT(16F5.0)
      CALL SUMMIT(FP,N,SUM)
      WRITE(6,20) SUM
   20 FORMAT(7H0SUM = ,F10.4)
      STOP
      END

      SUBROUTINE SUMMIT(A,M,TOT)
      DIMENSION A(M)
      TOT=0.0
      DO 10 I=1,M
C NOTE: LOCAL NAMES (USED WITHIN SUBROUTINES OR FUNCTIONS AND NOT
C TRANSFERRED VIA AN ARGUMENT LIST) ARE INDEPENDENT OF LOCATIONS WITH
C THE SAME NAMES USED IN OTHER PROGRAM MODULES (MAIN PROGRAMS,
C FUNCTIONS OR SUBROUTINES). THE SAME RULE APPLIES TO LABEL NUMBERS.
   10 TOT=TOT+A(I)
      RETURN
      END
```

43.

```
      WRITE(6,10)
      DO 1 N=3,20
      K=N+N-1
C NOTE: (K!/N!) IS EQUIVALENT TO K*(K-1)*.....*(N+1)
      N1=N+1
      A=FLOAT(N1)
      N2=N+2
      DO 2 I=N2,K
    2 A=A*FLOAT(I)
      N1=N-1
      B=1.0
      DO 3 I=2,N1
    3 B=B*FLOAT(I)
      FNO = (1.0/FLOAT(K)) * (A/B)
    1 WRITE(6,11) I, FNO
   10 FORMAT(14HONO OF SOURCES,14X,14HNO OF NETWORKS/)
   11 FORMAT(1H ,I5,13X,F15.0)
      STOP
      END
```

Output from this program

```
    NO OF SOURCES                NO OF NETWORKS

        3                          2.
        4                          5.
        5                         14.
        6                         42.
        7                        132.
        8                        429.
        9                       1430.
       10                       4862.
       11                      16796.
       12                      58786.
       13                     208012.
       14                     742900.
       15                    2674440.
       16                    9694847.
       17                   35357672.
       18                  129644808.
       19                  477638752.
       20                 1767263104.
```

44.

```
C THE DECLARATION 'REAL' ALLOWS YOU TO ALTER VARIABLE TYPE FROM
C INTEGER TO FLOATING-POINT. E.G. 'LYNX' WOULD NORMALLY BE INTEGER,
C BUT - AS IT HAS BEEN DECLARED 'REAL' - IT WILL BE FLOATING-POINT.
      REAL LYNX,LIMIT
C ARRAY SLINE IS USED IN PRINTING THE GRAPH
      DIMENSION SLINE(100)
C DATA STATEMENT SETS REQUIRED CHARACTERS
      DATA BLANK,LYNX,HARE,EDGE,TOP/1H ,1HY,1HX,1HI,1H-/
C ENSURE ARRAY SLINE HOLDS BLANK CHARACTERS
      DO 3 I=1,100
    3 SLINE(I)=BLANK
      WRITE(6,10)
   10 FORMAT(22H0PREY - PREDATOR MODEL/23H   YEAR   LYNX     HARE/)
C PRINT TOP BORDER OF GRAPH
      WRITE(6,12) EDGE,(TOP,I=1,100),EDGE
   12 FORMAT(1H ,24X,102A1)
C READ NYEARS (NO OF YEARS OF SIMULATION) AND LIMIT (UPPER LIMIT OF
C THE LYNX POPULATION). IF LIMIT IS ZERO OR NEGATIVE, NO UPPER LIMIT
C WILL BE SET. OTHERWISE, THE LYNX POPULATION WILL BE THE LESSER OF THE
C COMPUTED VALUE AND THE LIMITING VALUE.
      READ (5,13) NYEARS,LIMIT
   13 FORMAT(I3,F10.0)
C SET UP PARAMETERS
      X=100.
      Y=20.
      A=.93
      B=.04
      C=.004
      P=.22
      T=1.0/12.0
C OUTER LOOP = YEARS
      DO 1 I=1,NYEARS
C INNER LOOP=MONTHS
      DO 2 J=1,12
      D1=A*X-B*X*Y
      D2=C*X*Y-P*Y
C X = HARES, Y = LYNX
      X=X+D1*T
      Y=Y+D2*T
C CHECK THAT Y IS NOT GREATER THAN LIMIT
      IF(LIMIT.LE.0.0) GOTO 2
      Y=AMIN1(LIMIT,Y)
    2 CONTINUE
C K1 AND K2 GIVE PRINTER POSITIONS IN ARRAY SLINE FOR NO OF LYNXES AND
C NO OF HARES. AN UPPER LIMIT OF 150 (THOUSAND) HAS BEEN SET FOR THE
```

```
C GRAPH.  ANYTHING LARGER THAN THIS WILL BE SET TO 150 (THOUSAND)
      K1=INT(Y/150.0*99.0)+1
      K2=INT(X/150.0*99.0)+1
      IF(K1.GT.100) K1=100
      IF(K2.GT.100) K2=100
      SLINE(K1)=LYNX
      SLINE(K2)=HARE
C PRINT OUT INFORMATION FOR CURRENT YEAR
      WRITE(6,11) I, Y, X, EDGE, (SLINE(J),J=1,100), EDGE
   11 FORMAT(1H ,I4,2F9.0,2X,102A1)
C RESET SLINE TO BLANK CHARACTERS
      SLINE(K1)=BLANK
      SLINE(K2)=BLANK
    1 CONTINUE
C DRAW LOWER BOUNDARY OF GRAPH
      WRITE(6,12) EDGE,(TOP,I=1,100),EDGE
      STOP
      END
```

A sample printout from this program is shown overleaf.

Table D.14 (continued)

PREY – PREDATOR MODEL

YEAR	LYNX	HARE
1	24.	106.
2	25.	99.
3	25.	92.
4	25.	86.
5	25.	80.
6	25.	75.
7	25.	70.
8	25.	65.
9	25.	60.
10	25.	56.
11	25.	53.
12	25.	49.
13	24.	47.
14	23.	47.
15	22.	48.
16	22.	50.
17	22.	53.
18	22.	57.
19	22.	61.
20	22.	63.
21	23.	65.
22	24.	63.
23	25.	60.
24	25.	56.
25	25.	53.
26	25.	49.
27	24.	47.
28	23.	47.
29	22.	48.
30	22.	50.
31	22.	53.
32	22.	57.
33	22.	61.
34	22.	64.
35	23.	65.
36	24.	63.
37	25.	60.
38	25.	56.
39	25.	52.
40	25.	49.
41	24.	47.
42	23.	47.
43	22.	48.
44	22.	50.
45	22.	53.
46	22.	57.
47	22.	61.
48	22.	64.
49	23.	65.
50	24.	63.

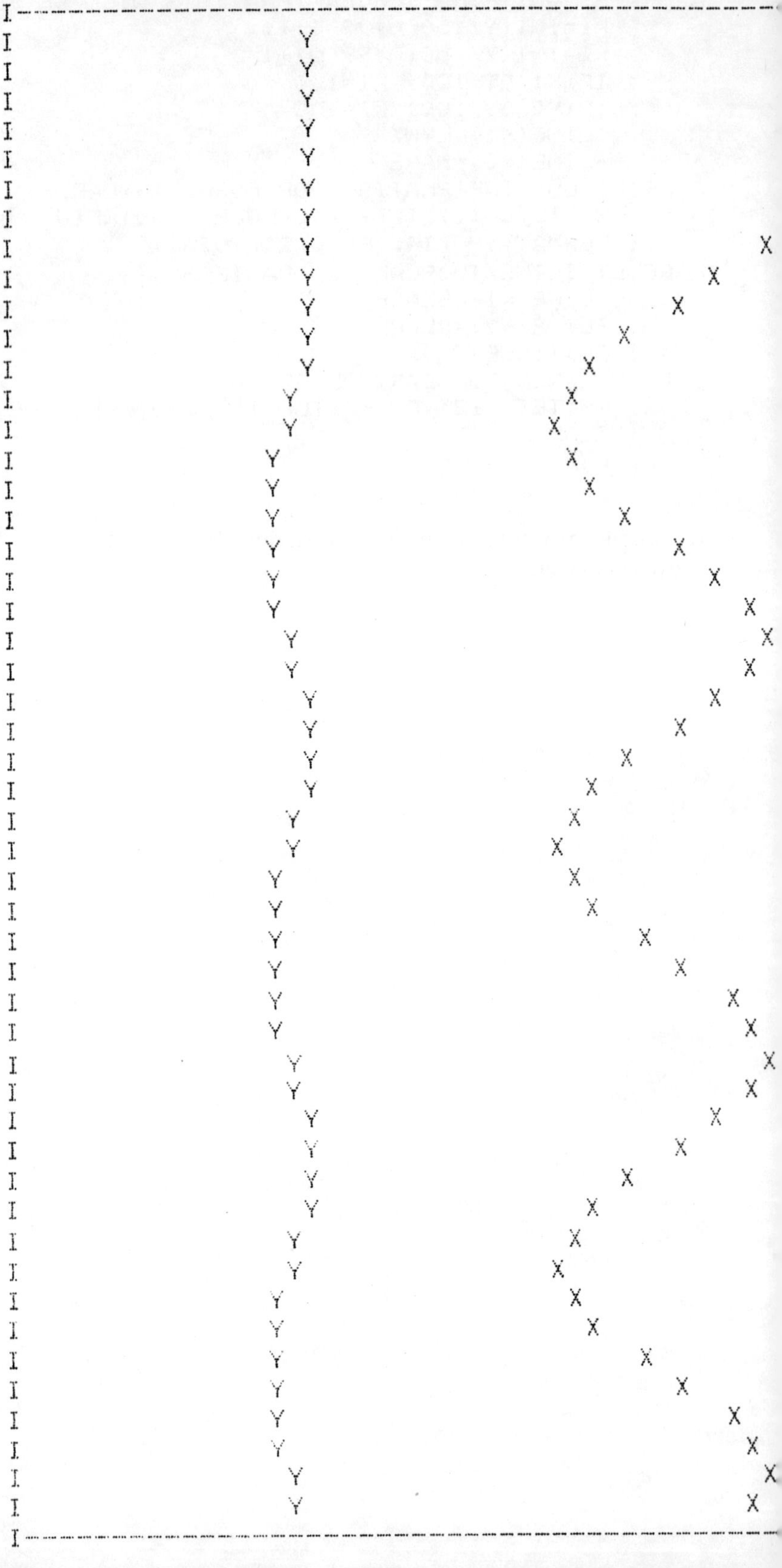

46. x = 28m, y = 24m, z = 24m. Profit is $244,000.

47.

```
      DIMENSION IX(20),SLINE(50)
      DATA STAR,BLANK,SLINE/1H*,1H ,50*1H /
      READ(5,1) N
    1 FORMAT(20I4)
      READ(5,1) (IX(I),I=1,N)
      MIN=IX(1)
      MAX=MIN
      DO 2 I=2,N
      IF(IX(I).GT.MAX) MAX=IX(I)
      IF(IX(I).LT.MIN) MIN=IX(I)
    2 CONTINUE
      RANGE=FLOAT(MAX-MIN)
      WRITE(6,10) MAX, MIN, RANGE
   10 FORMAT(7H0MAX = ,I4,2X,6HMIN = ,I4,2X,8HRANGE = ,F6.1//)
      FMIN=FLOAT(MIN)
      DO 3 I=1,N
      FP=FLOAT(IX(I))
    3 IX(I)=INT(((FP-FMIN)/RANGE)*49.0)+1
      DO 4 I=1,N
      K=IX(I)
      DO 5 J=1,K
    5 SLINE(J)=STAR
      WRITE(6,6) I, (SLINE(J),J=1,K)
    6 FORMAT(1H ,I4,2X,50A1)
      DO 4 J=1,K
    4 SLINE(J)=BLANK
      STOP
      END
```

```
MAX =   18  MIN =    7  RANGE =   11.0


   1  **************
   2  *
   3  ***********************
   4  **************************************************
```

51.

```
      DIMENSION X(20),PROB(20)
C MAXIMUM NUMBER OF X'S IS 20; INCREASE DIMENSION FOR MORE
      READ(5,1) N,N1,P,Q
```

```
C N IS NUMBER OF TRIALS AND N1 THE NUMBER OF ALTERNATIVE OUTCOMES
C EG: IN COIN-TOSSING EXAMPLE, N = 10 AND N1 = 11
    1 FORMAT(2I5,2F10.0)
      FN=N
      READ(5,2) (X(I),I=1,N1)
    2 FORMAT(16F5.0)
      FNFACT=1.0
      DO 3 I=2,N
    3 FNFACT=FNFACT*FLOAT(I)
C THE ARITHMETIC IS DONE IN FLOATING POINT AS THE LARGE NUMBERS
C GENERATED MAY BE TOO BIG FOR INTEGER STORAGE
      DO 4 I=1,N1
      XFACT=1.0
      IX=INT(X(I)+0.00001)
C THE USE OF INT(X+0.00001) WILL GIVE THE NEAREST INTEGER TO X
      IF(IX.LT.2) GOTO 6
      DO 5 J=2,IX
    5 XFACT=XFACT*FLOAT(J)
    6 NLESSX=N-IX
      FNX=1.0
      IF(NLESSX.LT.2) GOTO 4
      DO 7 J=2,NLESSX
    7 FNX=FNX*FLOAT(J)
    4 PROB(I)=(FNFACT/(XFACT*FNX))*(P**X(I))*(Q**(FN-X(I)))
      WRITE(6,8)
    8 FORMAT(27H0BINOMIAL LAW PROBABILITIES/11H          X ,2X,
     1 10H   PROB   //)
      WRITE(6,9) (X(I),PROB(I),I=1,N1)
    9 FORMAT(1H ,F10.1,2X,F10.5)
      STOP
      END
```

```
      BINOMIAL LAW PROBABILITIES
              X         PROB


            0.0       0.00098
            1.0       0.00977
            2.0       0.04395
            3.0       0.11719
            4.0       0.20508
            5.0       0.24609
            6.0       0.20508
            7.0       0.11719
            8.0       0.04395
            9.0       0.00977
           10.0       0.00098
```

52.

```
      DIMENSION X(30),PROB(30)
      PI=3.1415926
      READ(5,1) N
    1 FORMAT(I3)
      READ(5,2) XMEAN,VAR
    2 FORMAT(8F10.0)
      READ(5,2) (X(I),I=1,N)
      XMULT=1.0/SQRT(VAR*2.0*PI)
      DO 3 I=1,N
    3 PROB(I)=XMULT*EXP(-0.5*(X(I)-XMEAN)**2/VAR)
C EXP IS A FORTRAN LIBRARY FUNCTION RETURNING THE VALUE OF E TO THE
C POWER OF THE (FLOATING-POINT) ARGUMENT.
      WRITE(6,4) XMEAN, VAR
    4 FORMAT(33H0NORMAL DISTRIBUTION WITH MEAN = ,F10.4,17H  AND VARIANC
     1E = ,F10.4)
      WRITE(6,5)
    5 FORMAT(22H0       X          PROB  /)
      WRITE(6,6) (X(I),PROB(I),I=1,N)
    6 FORMAT(1H ,F10.4,2X,F10.4)
      STOP
      END
```

```
 NORMAL DISTRIBUTION WITH MEAN =     0.0000  AND VARIANCE =     1.

        X          PROB

     -5.0000      0.0000
     -4.0000      0.0001
     -3.0000      0.0044
     -2.0000      0.0540
     -1.9000      0.0656
     -1.2000      0.1942
     -1.0000      0.2420
     -0.5000      0.3521
      0.0000      0.3989
      0.5000      0.3521
      0.8000      0.2897
      1.0000      0.2420
      1.5000      0.1295
      2.0000      0.0540
      2.2800      0.0297
      2.5500      0.0154
      3.0000      0.0044
      3.5000      0.0009
      3.8000      0.0003
      5.0000      0.0000
```

53.

```
C            *** RUNS TEST ***
C SET UP ARRAY TO HOLD 1'S (+) AND 0'S (-)
      DIMENSION IRUNS(250)
C NO IS NUMBER OF PLUSSES AND MINUSES
      READ(5,1) NO
    1 FORMAT(I3)
C READ PLUSSES AND MINUSES CODED AS 1 (+) AND 0 (-)
      READ(5,2) (IRUNS(I),I=1,NO)
    2 FORMAT(80I1)
C N1 WILL COUNT +, N2 WILL COUNT -
      N1=0
      N2=0
C LAST REMEMBERS THE PREVIOUS VALUE (WHETHER + OR -)
      LAST=IRUNS(1)
C TEST WHETHER IRUNS(1) IS A + OR A -
      IF(LAST.EQ.1) N1=N1+1
      IF(LAST.EQ.0) N2=N2+1
      NRUN=1
C NRUN WILL COUNT THE NUMBER OF RUNS
      DO 3 I=2,NO
C TEST WHETHER IRUNS (I) IS A + OR A -
      IF(IRUNS(I).EQ.1) N1=N1+1
      IF(IRUNS(I).EQ.0) N2=N2+1
C TEST FOR END OF RUN. IF IRUNS(I) EQUALS LAST, THEN RUN IS NOT ENDED
      IF(IRUNS(I).EQ.LAST) GOTO 3
C IF RUN IS ENDED, STORE FIRST + OR - OF NEW RUN IN LAST AND INCREASE
C COUNT OF NO OF RUNS (NRUN) BY 1
      LAST=IRUNS(I)
      NRUN=NRUN+1
    3 CONTINUE
C OUTPUT COUNTS
      WRITE(6,4) N1, N2, NRUN
    4 FORMAT(6HON1 = ,I4,6H N2 = ,I4,14H NO OF RUNS = ,I4)
C IF EITHER N1 OR N2 < 20 TABLES WILL BE USED. OTHERWISE, THE Z VALUE
C IS COMPUTED
      IF(N1.GT.20) GOTO 6
      IF(N2.GT.20) GOTO 6
      WRITE(6,5)
    5 FORMAT(46HON1 OR N2 TOO SMALL FOR Z FORMULA - USE TABLES)
      STOP
C COMPUTE Z VALUE
    6 F1=FLOAT(N1*N2*2)
      F2=FLOAT(N1+N2)
      TOPLIN=FLOAT(NRUN)-((F1/F2)+1.0)
      BOTLIN=SQRT((F1*(F1-F2))/(F2*F2*(F2-1.0)))
```

```
  Z=ABS(TOPLIN/BOTLIN)
  WRITE(6,7) Z
7 FORMAT(11H0Z VALUE = ,F10.4)
  STOP
  END
```

APPENDIX E

Bibliography

Cole, J. P. (1969), 'Mathematics and Geography', *Geography, 54(2),* 152–164.

Cole, J. P. (1972), *Geography of world affairs.* Fourth edition. Penguin Books, Harmondsworth, Middlesex.

Cole, J. P. (1975), *Situations in human geography—a practical approach.* Blackwell, Oxford.

Cole, J. P. and King, C. A. M. (1968), *Quantitative geography.* Wiley, Chichester.

Doornkamp, J. C. and King, C. A. M. (1971), *Numerical methods in geomorphology.* Arnold, London.

Ebdon, D. S. (1976), *Statistics in geography—A practical approach.* Blackwell, Oxford.

Gregory, S. (1963), *Statistical Methods and the geographer.* Longmans, London.

Haggett, P. (1965), *Locational analysis in human geography.* Arnold, London.

Haggett, P. and Chorley, R. J. (1969), *Network analysis in geography.* Arnold, London.

Hammond, R. and McCullagh, P. S. (1974), *Quantitative techniques in geography—an introduction.* Clarendon Press, Oxford.

Harbaugh, J. W. and Bonham-Carter, G. (1970), *Computer simulation in geology.* Wiley–Interscience, New York.

Hollingdale, S. H. and Toothill, G. C. (1965), *Electronic computers.* Penguin Books, Harmondsworth, Middlesex.

King, L. J. (1969), *Statistical analysis in geography.* Prentice-Hall, Englewood Cliffs, New Jersey.

Leinbach, L. C. (1974), *Calculus with the computer.* Prentice-Hall, Inc., Englewood Cliffs, New Jersey.

Liao, L. H. and Scheidegger, A. E. (1968), A computer model for some branching-type phenomena in hydrology. *Bulletin of the International Association of Scientific Hydrology, 13(1),* 5–13.

Teach Yourself Operational Research. English Universities Press, London.

Turner, J. C. (1970), *Modern Applied Mathematics.* English Universities Press, London.